工程机械类专业"十三五"规划教材

工程机械专业英语

（第二版）

李美荣　主　编
贺玉斌　副主编

人民交通出版社股份有限公司
China Communications Press Co.,Ltd.

内 容 提 要

本书以工程机械的总体构造为基础,着重介绍工程机械的用途、构造、性能及应用,内容涵盖了发动机、底盘、电器设备以及路面施工机械、土方机械、压实机械和养护机械的相关专业词汇、专业知识和机械液电控制技术。本书共有三部分四单元,共二十七课。各篇内容相对独立,通过组合可以满足工程机械相关专业的教学需求。

本书是工程机械相关专业教学用书,可供公路机械化施工等专业教学使用,也可供工程技术人员学习参考。

图书在版编目(CIP)数据

工程机械专业英语/李美荣主编. —2 版. —北京:人民交通出版社股份有限公司,2018.1
ISBN 978-7-114-14375-5

Ⅰ.①工… Ⅱ.①李… Ⅲ.①工程机械—英语—职业教育—教材 Ⅳ.①TH2

中国版本图书馆 CIP 数据核字(2017)第 297116 号

Gongcheng Jixie Zhuanye Yingyu

书　　名:	工程机械专业英语(第二版)
著　作　者:	李美荣
责任编辑:	司昌静　富砚博
出版发行:	人民交通出版社股份有限公司
地　　址:	(100011)北京市朝阳区安定门外外馆斜街 3 号
网　　址:	http://www.ccpress.com.cn
销售电话:	(010)59757973
总 经 销:	人民交通出版社股份有限公司发行部
经　　销:	各地新华书店
印　　刷:	北京虎彩文化传播有限公司
开　　本:	787×1092　1/16
印　　张:	12.5
字　　数:	288 千
版　　次:	2008 年 6 月　第 1 版 2018 年 1 月　第 2 版
印　　次:	2023 年 6 月　第 2 版　第 4 次印刷
书　　号:	ISBN 978-7-114-14375-5
定　　价:	38.00 元

(有印刷、装订质量问题的图书由本公司负责调换)

前言

　　随着我国经济建设的高速发展，工程机械巨大的社会需求使我国成为国际工程机械的巨大市场，大大促进工程机械工业的快速发展，同时，也使得与工程机械相关的科研活动变得十分活跃。国外工程机械大量进入我国市场，对我国用户提出了新的要求，用户只有了解进口产品的结构、性能、正确使用，才能充分发挥产品的作业效率、延长其使用寿命。另外，随着我国工程机械工业的迅速崛起，出口机械产品的品种和数量也在迅速增加，相关科研成果也不断涌现。因此，准确阅读并翻译英文科技资料的重要性日益突出。在实际应用中，专业英语和公共英语存在着一定的距离，不懂专业的英语翻译将大大降低资料的准确性、专业性和应用性。本书的目的在于：①为工程机械专业的高职高专学生提供一个公共英语与专业英语融合的渠道，为学生实际工作中的应用打下基础；②（翻译部分）对青年教师讲授本课程提供一定的专业知识的指导；③对本领域的工程技术人员的学习提供一定的帮助。

　　本书内容难易适度，以工程机械的总体构造为基础，着重介绍工程机械的用途、构造、性能及应用。内容涵盖了发动机、底盘、电器设备以及路面施工机械、土方机械、压实机械和养护机械的相关专业词汇、专业知识和机械液电控制技术。

　　本书共有三部分四单元，共二十七课。各篇内容相对独立，通过组合可以满足工程机械专业的教学需求。

　　参加本书编写工作的有：内蒙古大学交通学院李美荣（第一部分12～16课及参考译文；第二部分1～11课、参考译文1～6课；第三部分及附录），内蒙古大学交通学院贺玉斌（第一部分1～11课），北奔重型汽车集团有限公司李轶楠（第二部分参考译文 7~11 课)。全书由李美荣任主编,贺玉斌任副主编。

  本书在编写过程中,得到交通系统相关院校领导和教师的大力支持,在此向大家表示感谢!

  由于我们经验有限,书中不妥和疏漏之处敬请读者指正,宝贵的意见请反馈至 teacherlili@ aliyun. com。

作　者

2017 年 8 月

Part 1 Engineering Machinery Bases

Unit 1　Texts ··· 2
　Lesson 1　Engine Construction and Operation Principles ·· 2
　Lesson 2　Compression-ignition Engines ··· 6
　Lesson 3　Diesel Engine Fuel Supply System ·· 8
　Lesson 4　Gasoline Engine Fuel Injection ·· 12
　Lesson 5　Engine Lubrication System ·· 16
　Lesson 6　Engine Cooling System ·· 20
　Lesson 7　Engine Starting System ·· 23
　Lesson 8　Gasoline Engine Ignition System ·· 26
　Lesson 9　Clutches ·· 30
　Lesson 10　Transmissions ··· 35
　Lesson 11　Torque Converters ··· 40
　Lesson 12　Differentials ·· 44
　Lesson 13　Final Drives ··· 46
　Lesson 14　Brake System ··· 49
　Lesson 15　Steering System ··· 52
　Lesson 16　Frame and Suspension System ··· 56
Unit 2　Reference Translations and Answers ··· 60
　第一课　发动机结构与工作原理 ·· 60
　第二课　压燃式发动机 ·· 62
　第三课　柴油发动机供油系统 ··· 63
　第四课　汽油发动机燃油喷射系统 ··· 65
　第五课　发动机润滑系统 ··· 66
　第六课　发动机冷却系统 ··· 68

第七课　发动机起动系统 …… 70
第八课　汽油发动机点火系统 …… 71
第九课　离合器 …… 73
第十课　变速器 …… 75
第十一课　液力变矩器 …… 78
第十二课　差速器 …… 80
第十三课　主减速器 …… 81
第十四课　制动系统 …… 83
第十五课　转向系统 …… 85
第十六课　车架和悬架系统 …… 87

Part 2　Engineering and Construction Machineries

Unit 1　Texts …… 90
Lesson 1　Bulldozers …… 90
Lesson 2　Scrapers …… 93
Lesson 3　Loaders …… 97
Lesson 4　Excavators …… 100
Lesson 5　Graders …… 104
Lesson 6　Asphalt Pavers …… 107
Lesson 7　Cement Concrete Pavers …… 111
Lesson 8　Rollers …… 115
Lesson 9　Asphalt Plants …… 120
Lesson 10　Cold Milling Machines …… 126
Lesson 11　Hot Recycling Machines …… 131

Unit 2　Reference Translations and Answers …… 136
第一课　推土机 …… 136
第二课　铲运机 …… 137
第三课　装载机 …… 139
第四课　挖掘机 …… 141
第五课　平地机 …… 143
第六课　沥青混凝土摊铺机 …… 145
第七课　水泥混凝土摊铺机 …… 147
第八课　压路机 …… 150
第九课　沥青混合料拌和设备 …… 152
第十课　冷铣刨机 …… 155
第十一课　热再生机 …… 159

Part 3　专业英语的翻译方法与技巧

第一节　专业英语特点 …………………………………………………………… 164
第二节　翻译方法和技巧 ………………………………………………………… 165

附录　工程机械常用英文缩写 …………………………………………………… 185
参考文献 …………………………………………………………………………… 192

>>>Part 1

Engineering Machinery Bases

◎ Unit 1   Texts

◎ Unit 2   Reference Translations and Answers

Unit 1　　Texts

Lesson 1

Engine Construction and Operation Principles

Internal-combustion engines are those engines that burn fuel inside the cylinders. Those that burn gasoline are known as gasoline engines. Other types of internal-combustion engines burn heavier oils or fuels. Of these types the diesel engine has come into the widest use. Each of this engine has a few main working parts; the auxiliary parts are necessary to hold the working parts together or to assist the main working parts in their performance.

The engines include two mechanisms and four(five) systems.

The two mechanisms are:

1. crank connecting rod mechanism, the parts are:

(1) engine block

(2) cylinder

(3) cylinder head

(4) piston

(5) connecting rod

(6) crankshaft

(7) flywheel

(8) crankcase

2. valve train mechanism, the parts are:

(1) inlet valve and exhaust valve

(2) valve lifter(tappet)

(3) push rod

(4) rocker arm

(5) camshaft

(6) valve timing gear

Those systems are:

(1) fuel supply system

(2) lubricating system

(3) cooling system
(4) starting system
(5) ignition system

All engines convert one form of energy into another. Internal-combustion engines convert the chemical energy stored in their fuels into heat energy during the burning part of their operation. The heat energy is converted into mechanical energy by the expansion of gases against pistons attached to crankshafts.

The engine transfer the energy resulting from gas expansion into useful work. The construction such as shown in Fig. 1-1-1.

The force exerted on the piston is transmitted through the connecting rod to the crankshaft that is made to revolve, ①turning through one half of a revolution as the piston moves downward. ②Attached to the crankshaft is a flywheel, which stores up energy. The momentum of the flywheel carries the piston through the balance of its motion until it receives another power impulse. In this way the reciprocating motion of the piston is transformed into a rotary motion at the crankshaft.

The number of strokes of the piston required to complete the cycle varies with the type of engine. The

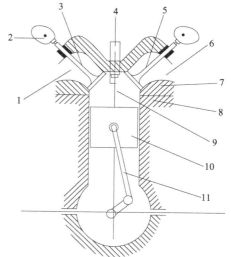

Fig. 1-1-1 Engine construction
1-intake port; 2-cam; 3-intake valve; 4-spark plug; 5-exhaust valve; 6-exhaust port; 7-cylinder head; 8-engine block; 9-combustion chamber; 10-piston; 11-connecting rod

cycle is generally extended through four strokes of the piston or two revolutions of the crankshaft. This is called a four-stroke-cycle or a four-cycle engine.

In the four-stroke-cycle engine, the four strokes are named suction, compression, power and exhaust.

Suction Stroke. During this stroke, the piston is moved downward by the crankshaft. Fresh mixture is sucked into the cylinder through the open inlet valve.

Compression Stroke. The compression, ignition, and much of the combustion of the charge take place during the next upward stroke of the piston. The mixture is ignited(fuel is injected into the fresh air in the diesel engine)③while under compression, and combustion is over half completed when the piston is at the top dead center.

It will be noted that both inlet and exhaust valves are closed during this stroke.

Power Stroke. The expansion of the gases due to the heat of combustion exerts a pressure in the cylinder and on the piston. Under this impulse the piston moves downward. Both valves also are closed during much of this stroke.

Exhaust Stroke. When the exhaust valve is opened, the greater part of burned gases escape because of their own expansion. The upward movement of the piston pushes the remaining gases out of the open exhaust valve.

New Words

1. convert [kən'vɜːt]　　　　　　　　　　　vt.　　　变换；转变
2. mechanical [mi'kænik(ə)l]　　　　　　 adj.　　机械的
3. expansion [ik'spænʃ(ə)n]　　　　　　　n.　　　膨胀
4. piston ['pist(ə)n]　　　　　　　　　　 n.　　　活塞
5. attach [ə'tætʃ]　　　　　　　　　　　　vt.；vi.　缚；附，连接
6. crankshaft ['kræŋkʃɑːft]　　　　　　　n.　　　曲轴
7. construction [kən'strʌkʃ(ə)n]　　　　n.　　　结构
8. transmit [trænz'mit; trɑːnz-; -ns-]　　vt.　　　传送；传导，发射
9. revolve [ri'vɔlv]　　　　　　　　　　　vi.　　　旋转
10. revolution [revə'luːʃ(ə)n]　　　　　　n.　　　转（速），革命
11. flywheel ['flaiwiːl]　　　　　　　　　n.　　　飞轮
12. momentum [mə'mentəm]　　　　　　n.　　　动量；冲量
　　　　　　　　　　　　　　　　　　　　　　　　（复数 momentums 或 momenta）
13. impulse ['impʌls]　　　　　　　　　　n.　　　推动，脉冲，冲量
　　　　　　　　　　　　　　　　　　　　vt.　　　推动
14. transform [træns'fɔːm; trɑːns-; -nz-] vt.；vi.　改变；转变；变形
　　　　　　　　　　　　　　　　　　　　n.　　　改变
15. reciprocate [ri'siprəkeit]　　　　　　vt.；vi.　往复，互换
16. rotary ['rəut(ə)ri]　　　　　　　　　 adj.　　旋转的
17. stroke [strəuk]　　　　　　　　　　　n.　　　冲程；行程，打；击
18. extend [ik'stend; ek-]　　　　　　　 vt.；vi.　延续，伸展，延长
19. suction ['sʌkʃ(ə)n]　　　　　　　　　n.　　　吸入；吸收
20. compression [kəm'preʃ(ə)n]　　　　 n.　　　压缩；压紧
21. exhaust [ig'zɔːst; eg-]　　　　　　　n.；vt.　排出；排气，用尽
22. inlet ['inlet]　　　　　　　　　　　　n.　　　入口；进口，引入
23. valve [vælv]　　　　　　　　　　　　n.　　　气门；阀门
24. ignition [ig'niʃ(ə)n]　　　　　　　　n.　　　点火
25. charge [tʃɑːdʒ]　　　　　　　　　　　n.　　　负荷，货物，充电；
　　　　　　　　　　　　　　　　　　　　vt.　　　充电，装料；控告；索价
26. ignite [ig'nait]　　　　　　　　　　　vt.；vi.　点火；着火
27. escape [i'skeip; e-]　　　　　　　　vt.；vi.　逃脱
28. half-completed　　　　　　　　　　　adj.　　完成一半的
29. remain [ri'mein]　　　　　　　　　　vi.　　　剩下，留下，保持，仍是

Phrases and Expressions

1. (be) attached to...　　　　　　　　　　　　　　　　连接(安装于)……；附属于……，依恋

2. ...such as shown... 如图所示的……
3. connecting rod 连杆
4. four-stroke-cycle-engine 四冲程循环发动机
5. inlet valve 进气门
6. exhaust valve 排气门
7. top-dead-center 上止点
8. bottom dead center 下止点

Notes

①...,turning through one-half of a revolution as the piston moves downward.
当活塞向下运动(一次),曲轴转半圈。
turning...含有时间状语从句的分词短语,作状语。
②Attached...被动结构的过去分词放在句首,形成倒装语序。
③...while under compression = ...while it is under compression

Exercise One

Put the following expressions into Chinese:
1. internal combustion engine
2. liquid-cooled engine
3. power stroke
4. crankshaft
5. cylinder wall wear
6. the camshaft

Exercise Two

Choose one of the four words or phrases that best matches the definition according to the text.
1. The _____ is main shaft of the engine and in conjunction with connecting rods, changes reciprocating motion of piston into rotary motion. ( )
   a. differential shaft　　　　　b. final drive shaft
   c. crankshaft　　　　　　　　d. camshaft
2. The _____ is a heavy wheel which absorbs and stores of the energy by means of momentum.
   a. gearing wheel　　　　　　b. sprocket wheel
   c. automobile tyre　　　　　 d. flywheel
3. The _____ is a shaft containing lobes or cams which operate engine valves.
   a. revolving shaft　　　　　　b. camshaft
   c. crankshaft　　　　　　　　d. driving shaft
4. The _____ permits a fluid or gas to enter a chamber and seals against exit.
   a. by-pass valve　　　　　　b. differential valve
   c. intake valve　　　　　　　d. exhaust valve

Exercise Three

Answer the questions:
1. Which strokes does the four stroke engine include?

2. What is the purpose of the engine?

lesson 2

Compression-ignition Engines

The four-stroke compression ignition engine has a compression ratio of 16-24 to 1 and raises an induced charge of air to a pressure of 3-5①MPa and a temperature of 500-800℃ at the end of the compression stroke. The fuel used has a self-ignition temperature of some 400℃ and therefore ignites when injected into the combustion chamber at an injection pressure of 10-25MPa.

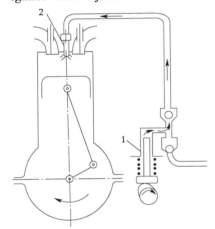

Fig.1-2-1　Compression-ignition engine construction
1-injection pump;2-injector

In the petrol engine a homogeneous combustible mixture, supplied by the carburetter, is ignited by the sparking plug, but on the compression-ignition engine(Fig.1-2-1) the thorough mixing together of the fuel and air for complete combustion must be accomplished within the cylinder during some 35°-40° of crankshaft revolution. This presents difficulty and is accomplished by using a penetrating spray of finely-divided fuel and considerable turbulence of the compressed air in the cylinder. Even so, only some 80% of air supplied can be utilized, but the spark-ignition engine can consume all the air induced.

Compared with the spark-ignition engine the high combustion pressures need a strong, heavy and costly construction,② while the fine limits to which the fuel pump and injectors must be manufactured add to the expense. Even with extensive light alloy construction, the power to weight ratio is lower than for a petrol engine. Starting may present more difficulty due to the high compression pressure and the heat required for fuel ignition. Idling is less smooth, acceleration slower, maximum③ rev/min are lower and imperfect combustion will result in a lower④ b.m.e.p and obnoxious exhaust smoke.

On the other hand, the higher compression ratio gives a greater thermal efficiency and thus an increased economy⑤km/L. The fuel is cheaper and has a much reduced fire risk. Engine torque is greater at low engine speeds and in practice considerably less maintenance is required. Carbon, beyond a thin coating, does not form when the engine is operating correctly.

New Words

1. spark[spɑːk]　　　　　　　　　n.　　　火花
   　　　　　　　　　　　　　　　vt.;vi　　发火;激发
2. ratio['reiʃiəu]　　　　　　　　n.　　　比;比率;比例
3. inject[in'dʒekt]　　　　　　　vt.　　　喷射;射入
4. injection[in'dʒekʃ(ə)n]　　　n.　　　注射,喷射,注射液

Part 1 Engineering Machinery Bases

5. injector [inˈdʒektə] n. 注射器;喷油器
6. chamber [ˈtʃeimbə] n. 房间,燃烧室,会所
7. homogeneous [ˌhɔmə(u)ˈdʒi:niəs] adj. 同样的,均匀的
8. carburetter [ˈkɑ:bjuretə] n. 化油器;汽化器
   (carbureter, carburettor, carburetor)
9. plug [plʌg] n. 塞子,火花塞,插头
10. penetrating [ˈpenitreitiŋ] adj. 穿透的,渗透的
11. spray [sprei] n. 雾,喷雾,喷嘴,浪花
     vt.;vi. 喷雾
12. divide [diˈvaid] vt. 分;分开,分配,除
13. considerable [kənˈsid(ə)rəb(ə)l] adj. 可观的,相当的,不可忽视的
     considerably adv.
14. turbulence [ˈtə:bjul(ə)ns] n. 涡流,湍流;骚动
15. limit [ˈlimit] n. 限度,极限;公差
16. alloy [ˈælɔi] n. 合金
17. idling [aidliŋ] n. 空转;惰转;慢车;怠速
18. acceleration [əkseləˈreiʃ(ə)n] n. 加速;加速度
19. imperfect [imˈpə:fikt] adj. 不完全的,不完美的
20. thermal [ˈθə:m(ə)l] adj. 热的,温热的
21. risk [risk] n. 危险,危险性
22. torque [tɔ:k] n. 转矩,扭矩
23. maintenance [ˈmeint(ə)nəns] n. 维修,保养;维持
24. carbon [ˈkɑ:b(ə)n] n. 碳,积炭
25. coating [ˈkəutiŋ] n. 覆盖层

Phrases and Expressions

1. compression-ignition engine   压燃式发动机
2. spark-ignition engine   火花点燃式发动机
3. combustion chamber   燃烧室
4. the ratio of ... to ...   比……
5. sparking plug   火花塞
6. finely-divided   喷(分)得很细的
7. light-alloy   轻合金的

Notes

①MPa 兆帕(压力单位)
②... while the fine limits to which the fuel pump and injectors must be manufactured add to

the expense.

……同时,油泵和喷油器必须加工很精密,这也增加了它的成本。

while 连接词,引出时间状语从句。

③rev/min = revolutions per minute   转/分

④b. m. e. p = brake mean effective pressure

制动平均有效压力(动力性指标)

⑤km/L = kilometer(s) per litre   每升公里数(经济性指标)

Exercise One

Put the following expressions into Chinese.

1. oil control ring
2. compression ring
3. compression ignition engine
4. spark ignition engine
5. crankpin
6. valve train mechanism

Exercise Two

Choose one of the four words that best matches the definition according to the text.

1. The diesel engines include _____ systems.
   a. one   b. two   c. three   d. four
2. The ignition type of diesel engine is the _____.
   a. compression ignition   b. spark ignition
   c. compression and spark combination ignition   d. detonation
3. Compared with gasoline engines, the diesel engines need _____ construction.
   a. lighter   b. stronger
   c. lower cost   d. cheaper
4. Compared with gasoline engines, the diesel engines have _____.
   a. higher compression ratio   b. lower compression ratio
   c. the same ratio   d. uncertainty

Exercise Three

Answer these questions:

1. What are the two type of piston rings?
2. What is the purpose of the connecting rod?

Lesson 3

Diesel Engine Fuel Supply System

The type of fuel available for use in diesel engine varies from highly volatile jet fuel and kerosene, to the heavier furnace oil. How well a diesel engine can operate with different types of fuel is dependent upon engine operation conditions, as well as fuel characteristics.

Diesel fuel is a mixture of kerosene, gas oil and solar oil fractions obtained after distillation of gasoline fraction from petroleum. The main characteristics of diesel fuel are ignitability estimated in cetane numbers, viscosity, pour point, purity, etc. Diesel fuel is produced in different grades, which differ mainly in the pour points, flashing points and viscosity values.

A diesel engine fuel system consists of a fuel tank, a primary filter, a secondary filter, a fuel supply pump with a hand primer, an injector pump with a speed governor and automatic injection timing clutch, nozzle holders with nozzles, low and high pressure fuel lines(Fig. 1-3-1).

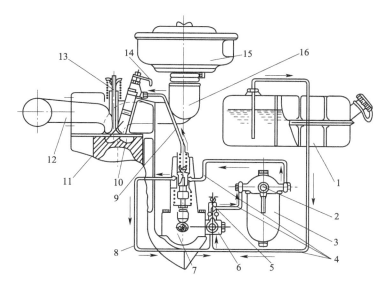

Fig. 1-3-1 Diesel engine fuel supply system
1-fuel tank; 2-relief valve; 3-diesel filter; 4-low pressure pipes; 5-hand primer pump; 6-delivery pump; 7-fuel injecting pump; 8-returning pipes; 9-high pressure pipes; 10-combustion chamber; 11-fuel injector; 12-exhaust pipe; 13-exhaust valve; 14-drain pipes; 15-air cleaner; 16-intake manifold

During operation of the engine, the fuel supply pump draws fuel from the tank, forces it through the primary filter and delivers through the secondary filter to the injector pump. From the injector pump the fuel is fed through the high pressure line to nozzles, the fuel atomized by the nozzles is injected into the cylinders according to the engine firing order. Surplus fuel is returned from the injection pump and nozzles to the fuel tank. The air is supplied to the cylinders through the air cleaner.

The fuel injector pump is intended to inject fuel under high pressure to the engine cylinders in a particular sequence. The injector pump is disposed between the cylinder banks and is driven from the camshaft by means of a gear train. The pump comprises a body, a camshaft, the sections (according to the number of cylinders), and a plunger control mechanism. ①The front part of the injector pump carries a positive speed governor which meters the fuel in accordance with the load thus maintaining the engine speed preset by the driver. The rear end of the pump camshaft mounts an injection timing clutch which is used to change the instant of fuel injection depending on the engine speed.

A section of the injector pump consists of a plunger and a barrel, a roller tappet and a delivery valve. The barrel has two ports located at different levels, the plunger top is also provided with two ports and a helix. The plunger is lapped to the barrel. ②When the plunger moves down forced by the spring, the fuel under a slight pressure created by the fuel supply pump flows through the longitudinal inlet passage in the body filling the space above the plunger. As the plunger is moved upward by the cam and tappet, the fuel is bypassed to fuel passage till the plunger edge seals off the barrel port. As the plunger continues to move upward the pressure reaches the delivery valve limit, the plunger lifts slightly and the fuel is discharged through the high-pressure line to the nozzle. The plunger keeping on moving builds up a pressure that overcomes the nozzle stem spring load. The stem lifts up and the fuel injection begins. The injection continues until the edge of the plunger helix opens the port in the barrel; now the fuel pressure drops, the relief band of the delivery valve is lowered to the seat by the spring thus increasing the volume in the fuel line between the nozzle and valve, hence ensuring positive shutoff of the fuel. When the rack is moved, the plunger rotates and the helix edge opens the barrel port in advance or with delay so that the time of port opening and quantity of fuel injected into the cylinder are changed.

The nozzle is used to inject metered quantities of finely atomized fuel under pressure into the cylinders.

The nozzle should be optimized for the particular engine type. This means optimum combustion, minimal pollution emissions and full engine output. The closed-type nozzles include pintle nozzle and hole-type nozzle. Pintle nozzles are used with in-line injection pumps on indirect-injection engines. Hole-type nozzles are used with in-line injection pumps on direct-injection engines.

The closed-type nozzle consists of a steel nozzle holder, a cap nut, a spray nozzle, a stem (or needle), a spindle and a filter. The fuel passes through the filter, a vertical passage, annular slot to the fuel space of the spray nozzle. When the pressure in the fuel space overcomes the spring, the stem is lifted from the seat and the fuel is injected into the combustion chamber. As the pressure in the fuel line drops, the stem is shut off. Surplus fuel is by passed via a return line to the tank.

New Words

1. volatile ['vɔlətail]   n.   挥发性,易变,反复无常
2. kerosene ['kerəsiːn]   n.   煤油
   (kerosine)
3. fraction ['frækʃ(ə)n]   n.   分馏物,馏分
4. petroleum [pə'trəuliəm]   n.   石油;石油产品
5. distillation [ˌdisti'leiʃn]   n.   蒸馏,蒸馏液
6. ignitability [igˌnaitə'biləti]   n.   可燃性;着火性
7. cetane ['siːtein]   n.   十六烷值

Part 1 Engineering Machinery Bases

8. viscosity[vi'kɔsiti]	n.		黏度；黏性；黏滞性，内摩擦
9. primer[praim, prim]	n.		注油器；起动注油器
10. injector[in'dʒektə]	n.		喷油器；喷射器
11. governor['gʌv(ə)nə]	n.		调节器；调速器
12. nozzle['nɔz(ə)l]	n.		喷嘴；喷口
13. atomize['ætəmaiz]	vt.		使雾化；把…喷成雾状
14. surplus['sə:pləs]	adj.		过剩的；剩余的
15. plunger['plʌn(d)ʒə]	n.		柱塞
16. barrel['bær(ə)l]	n.		套筒，筒体
17. tappet['tæpit]	n.		挺杆，挺柱
18. helix['hi:liks]	n.		螺旋槽，螺旋结构
19. longitudinal[,lɔn(d)ʒi'tju:din(ə)l]	adj.		纵向的，轴向的
20. by-pass[bai pɑ:s]	n.		旁通管；旁路
21. edge[edʒ]	n.		刃口；刀口，边
22. stem[stem]	n.		柄；杆，堵塞物
	vt.		给……装柄(杆，把)
23. annular['ænjulə]	adj.		环形的，有花纹的
24. slot[slɔt]	n.		切口，槽，沟，轨迹
25. spindle['spind(ə)l]	n.		轴；轴颈

Phrases and Expressions

1. vary from... to	从……到不等(变化)
2. furnace oil	燃料油；炉用燃油
3. solar oil	太阳油；灯用煤油
4. cetane number	十六烷值
5. nozzle holder	喷油器体
6. in accordance with	按照；根据
7. delivery valve	出油阀，输油阀
8. relief band	减压环带
9. in advance	提前；预先；预付
10. spray nozzle	喷油嘴

Notes

①The front part of the injector pump carries a positive speed governor which meters the fuel in accordance with the load thus maintaining the engine speed preset by the driver.
喷油泵的前端装有一个调速器，根据负荷大小供给燃油，从而保持驾驶员预定的发动机速度。

②When the plunger moves down forced by the spring, the fuel under a slight pressure created by the fuel supply pump flows through the longitudinal inlet passage in the body filling the space above the plunger.

当柱塞由弹簧强制下移时，燃油在燃油泵产生的低压作用下经纵向进油通道进入柱塞上部的泵腔。

Exercise One

Put the following expressions into Chinese.

1. low and high pressure fuel lines
2. plunger and barrel assembly
3. fuel injector pump
4. fuel supply pump
5. hole-type spary nozzle
6. pintle-type spary nozzle

Exercise Two

Choose one of the four words or phrases that best matches the definition: according to the text

1. The _____ is a process of changing from a liquid to a vapour.
   a. evaporation   b. expansion
   c. ventilation   d. radiation
2. The _____ connects engine to muffler to conduct spent gases away from engine.
   a. propeller shaft b. differential gear
   c. exhaust pipe   d. suction line
3. The _____ injects or inserts a fluid or gas, usually against pressure, into a cylinder or chamber.
   a. distributor   b. injector
   c. carburetor   d. accelerator
4. The _____ is a pressure less than atmospheric pressure.
   a. vacuum   b. charge
   c. exhaust   d. suction

Exercise Three

Answer the questions:

1. What are the main characteristics of the diesel fuel?
2. What components does the diesel engine fuel supply system include?

Lesson 4

Gasoline Engine Fuel Injection

There are two types of fuel injection system used with comprehensive computerized engine control systems single-point and multipoint. They each use an intermittent or timed spray to control fuel quantity.

Single point injection is often referred to as throttle body injection. Single point means that fuel is introduced into engine from one location. This system uses an intake manifold similar to what would be used with a carbureted engine, but the carburetor is replaced with a throttle body unit.

The throttle body unit contains one or two solenoid-operated injectors that spray fuel directly over the throttle blade (or blades) (Fig. 1-4-1). Fuel under pressure is supplied to the injector. The throttle blade is controlled by the throttle linkage just as in a carburetor. The computer controls voltage pulses to the solenoid-operated injector, which opens and sprays fuel into the throttle bore for a certain length of time depending on engine conditions relayed by sensors. The longer that the injector is open, the more fuel is injected. As engine load and rpm are increased, the injector open times are increased to match increasing airflow. The length of time is referred to as pulse width. The longer the pulse width, the more fuel is injected. The amount of air introduced is controlled by the opening of the throttle blade. The sensors commonly are rpm, airflow, manifold pressure, throttle position, water temperature, air temperature and oxygen sensors.

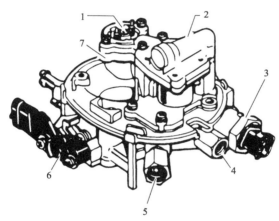

Fig. 1-4-1 Throttle body injection unit

1-fuel injector; 2-fuel meter cover; 3-idle control valve; 4-fuel return nut (to tank supply); 5-fuel inlet nut (from fuel pump and tank supply); 6-throttle position sensor; 7-fuel meter body

The EFI system is characterized by excellent throttle response and good driveability. [1]Experience has shown, however, that the system is best suited for engines with small cross section area manifold runners that at low speeds will keep the fuel mixture moving at a higher velocity. This reduces the tendency for the heavier fuel particles to fall out of the airstream.

Multipoint injection is often referred to as port injection and means that fuel is introduced into the engine from more than one location. This system uses an injector at each intake port. Fuel is sprayed directly into the port, just on the manifold side of the intake valve.

The multipoint injection system provides the most advanced form of fuel control yet developed. It offers the following advantages:

1) [2]Spraying precisely the same amount of fuel directly into the intake port of each cylinder

eliminates the unequal fuel distribution so inherent when already mixed air and fuel are passed through an intake manifold.

2) Because there is no concern about fuel condensing as it passes through the intake manifold, there is less need to heat air or the manifold.

3) Because there is no concern about fuel molecules falling out of the airstream while moving through the manifold at low speeds, the cross-section area of the manifold runners can be larger and thus offer better cylinder-filling ability (volumetric efficiency, VE) at higher speeds.

4) Most of the manifold-wetting process is avoided; some wetting still occurs in the port areas. If fuel is introduced into the intake manifold, some will remain on the manifold floor and walls especially during cold engine operation and acceleration. Fuel metering has to allow for this fuel in order to avoid an overlean in the cylinders. It has to be accounted for again during high-vacuum conditions because it will then begin to evaporate and go into the cylinders.

In general port fuel injection provides better engine performance and excellent driveability while maintaining or lowering exhaust emission levels and increasing fuel economy.

New Words

1. comprehensive [kɔmpri'hensiv]   adj.   理解的,综合的,广泛的
2. computerize [kəm'pju:təraiz]   vt.   使电子计算机化
3. multipoint ['mʌltipɔint]   n.   多点式
4. intermittent [intə'mit(ə)nt]   adj.   间断的,周期性的
5. quantity ['kwɔntiti]   n.   量;大量,数量
6. refer [ri'fə:]   vt.;vi.   把……归诸;归因,参考,涉及
7. carburet ['kɑ:bjuret]   vt.   使气体与汽油混合,使与碳化合
8. solenoid ['səulənɔid]   n.   螺线管;圆筒形线圈
9. energize ['enədʒaiz]   vt.;vi.   增能,通电,活动
10. characterize ['kærəktəraiz]   vt.   表示……的特征
11. experience [ik'spiəriəns]   vt.;vi.;n.   经验;体验
12. velocity [və'lɔsəti]   n.   速度
13. tendency ['tend(ə)nsi]   n.   倾向,趋势
14. particle ['pɑ:tik(ə)l]   n.   颗粒;微粒;粒子
15. airstream ['eəstri:m]   n.   气流;空气射流
16. eliminate [i'limineit]   vt.   消除;消去
17. distribution [distri'bju:ʃ(ə)n]   n.   分配;分布
18. inherent [in'hiər(ə)nt; -'her(ə)nt]   adj.   固有的;内在的
19. concern [kən'sə:n]   vt.;n.   涉及;和……有关系
20. condense [kən'dens]   vt.;vi.   冷凝
21. molecule ['mɔlikju:l]   n.   分子,微小颗粒
22. evaporate [i'væpəreit]   vt.;vi.   蒸发,使蒸发

Part 1　Engineering Machinery Bases

Phrases and Expressions

1. single-point injection 　　　　　　　单点喷射
2. multi-point injection 　　　　　　　多点喷射
3. be referred to as 　　　　　　　　　指……
4. solenoid-operated 　　　　　　　　电磁线圈控制的,电磁操纵的
5. be characterized by 　　　　　　　以……为特点
6. cylinder-filling 　　　　　　　　　　汽缸充量
7. volumetric efficiency(VE) 　　　　容积效率
8. manifold-wetting 　　　　　　　　进气歧管浸油
9. throttle body injection(TBI) 　　　(化油器体)节气门段喷射
10. electronic fuel injection(EFI) 　　电子燃油喷射
11. throttle response 　　　　　　　　(发动机)对节气门开度的反应灵敏度

Notes

①Experience has shown, however, that the system is best suited for engines with small cross-section area manifold runners that at low speeds will keep the fuel mixture moving at a higher velocity.

但是,实践证明该系统最适用于进气歧管横截面小的发动机,这样发动机低速时仍能保持可燃混合气在歧管内高速流动。

②Spraying precisely the same amount of fuel directly into the intake port of each cylinder eliminates the unequal fuel distribution so inherent when already mixed air and fuel are passed through an intake manifold.

精确地将等量的燃油直接喷入每个汽缸的进气口,消除了原来已混合的空气和燃油通过进气管时燃油本质上分配不均的现象。

spraying 分词短语作句子的主语。

Exercise One

Put the following expressions into Chinese.

1. compressed mixture
2. crankshaft position sensor
3. electric fuel pump
4. electric discharge in the form of a spark
5. volumetric efficiency
6. high tension current

Exercise Two

Choose one of the four words or phrases that best matches the definition according to the text.

1. The _____ surges necessary to produce the sparks at the spark plugs are provided by the ignition system components.

a. low voltage   b. high voltage   c. 12v   d. 24v

2. At high speeds, the current flow to the primary coil has only a few milliseconds to flow due to the rapid opening and closing of the _____.

a. distributor   b. condense   c. contact point   d. battery

3. The resistor is in series between the battery and the _____ in order to keep primary circuit voltage at a desired level.

a. primary coil winding   b. secondary coil winding

c. breaker   d. distributor

4. The magnetic field produced by the primary coil winding must completely saturate the _____.

a. primary coil winding   b. secondary coil winding

c. ignition coil   d. mutual coil

Exercise Three

Answer the questions:

1. Which types does the gasoline engine fuel injection system include?

2. What is the main characteristic of the EFI system?

Lesson 5

Engine Lubrication System

Without the aid of friction, an automobile could not move itself. Excessive friction in the engine, however, would mean rapid destruction. Internal friction cannot be eliminated, but it can be reduced to a considerable degree by the use of friction reducing lubricants.

Lubricating oil in an automobile engine has several tasks to perform:

(1) By lubrication, reduce friction between moving parts of the engine.

a) Reduce amount of destructive heat generated by excessive friction;

b) Conserve power that would otherwise be wasted in overcoming excessive friction.

(2) By acting as a seal to prevent leakage between parts such as pistons, rings and cylinders.

(3) By flowing between friction-generating parts to carry away heat.

(4) By washing away abrasive metal worn from friction surfaces.

In modern automobile the lubrication system is classified as "pressure" or "splash", [①]although various combination of these two systems are used. Most passenger cars use the pressure system in which the oil is forced under pressure by a gear pump to most of the various rotating or reciprocating parts.

The splash system utilizes dippers on the ends of the connecting rods to splash the oil on the various parts as they travel through oil troughs. Splashed oil usually becomes a mist for lubricating parts. A pump is employed to carry the oil to the troughs.

In the pressure system oil is supplied by an oil pump driven by the camshaft. The pump sucks the oil from the sump to an oil gallery or header that runs the length[2] of the block. Most oil galleries are holes drilled in the block, although some engines use steel or copper tubes. The header supplies oil directly to the main bearing through connecting passages. From the main bearings it is forced through holes drilled in the crankshaft to the connecting rod bearings. From the connecting rod the lubricant is forced through holes drilled in the connecting rods to the wristpins. Leads from the main bearings deliver oil under pressure from the main bearings to the camshaft bearings. One of these leads also supplies an oil jet for lubricating the timing gears. In the design shown, oil spray holes on the upper part of the connecting rod bearing are used to lubricate the thrust side of the cylinder wall.

Overhead valve engines such as shown in Fig. 1-5-1 provide for valve-operating mechanism lubrication by connecting a vertical header to the engine main header to feed oil to the hollow rocker-arm shaft. The shaft generally contains metering devices at each valve to supply the required amount of oil. Oil passages[3] in the head allow the oil to return to the sump by gravity. Even in the full-pressure system, the oil from the ends of the connecting rod bearings is thrown up to lubricate the cylinder walls and cams. The oil rings on the pistons control this distribution, carrying the oil upward to the cylinder walls and wiping and draining off excess oil. Excess oil escaping from the bearing ends and falling from pistons, cylinders and so on, of course, is returned to the oil sump for re-circulation.

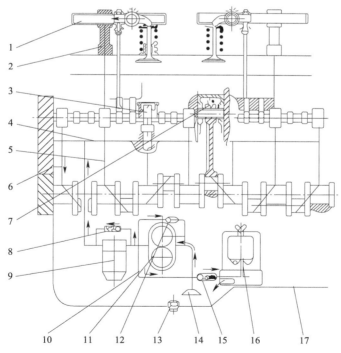

Fig. 1-5-1 Lubricating system

1-rocker arm shaft; 2-vertical header; 3-oil pump drive shaft; 4-oil header; 5-oil passage; 6-nozzle; 7-connecting rod passage; 8-by pass valve of primary oil filter; 9-primary oil filter; 10-oil pipes; 11-oil pump; 12-pressure limiting valve; 13-drain tap; 14-collecting filter; 15-pressure limiting valve of secondary oil filter; 16-secondary oil filter; 17- oil sump

Oils for engines fall into two basic categories: petroleum-based oils and synthetic oils. Petroleum-based oils contain a variety of additives, so in the fact they are partly synthetic. Antiscuff additives help to polish moving parts, including cams, pistons and cylinder walls. This is particularly important during new engine break-in and after an engine overhaul. Corrosion inhibitors reduce the formation of harmful acids by attacking the acid-forming ingredients. Corrosion inhibitors neutralize combustion by-products (acids and water) before they can do any harm to the engine. Detergent-dispersants clean engine parts during operation and keep these contaminations in suspension in the oil. Viscosity index improvers tend to stabilize or improve the viscosity of engine oils at various temperature, they improve the body and fluidity of the oil. Foam inhibitors reduce the tendency of oil to foam. Heat and agitation mix the oil with air to create foam. Oil foam reduces the lubricating ability of the oil and causes oil starvation and failure of engine parts.

New Words

1. lubrication [ˌluːbriˈkeiʃən]　　　　　　n.　　　润滑;润滑作用
2. destruction [diˈstrʌkʃ(ə)n]　　　　　　n.　　　破坏;毁坏,消灭
3. lubricant [ˈluːbrik(ə)nt]　　　　　　　n.　　　润滑剂;润滑油
4. destructive [diˈstrʌktiv]　　　　　　 adj.　　 破坏性的;有害的
5. abrasive [əˈbreisiv]　　　　　　　　 n.　　　磨料,磨蚀剂
6. dipper [ˈdipə]　　　　　　　　　　　n.　　　勺,铲斗,挖掘机
7. trough [trɔf]　　　　　　　　　　　　n.　　　槽;油槽
8. sump [sʌmp]　　　　　　　　　　　n.　　　油池,油底壳;集油槽
9. gallery [ˈɡæl(ə)ri]　　　　　　　　　n.　　　油道,通道,走廊;画廊
10. copper [ˈkɔpə]　　　　　　　　　　n.　　　铜
11. hole [həul]　　　　　　　　　　　　n.　　　孔,洞
12. drill [dril]　　　　　　　　　　　 vt.;vi.　　钻,钻孔,训练
13. bearing [ˈbeəriŋ]　　　　　　　　　n.　　　轴承,支承
14. wristpin [ˈristpin]　　　　　　　　　n.　　　活塞销,肘节,(曲柄,十字头,偏心轮)销
15. lead [liːd]　　　　　　　　　　　　n.　　　导管,油路,分油道,领导
16. thrust [θrʌst]　　　　　　　　　　　n.　　　推力
　　　　　　　　　　　　　　　　　 vt.;vi.　　推力,冲
17. overhead [ˌəuvəˈhed]　　　　　　　adj.　　 上面的;头上的,顶置
　　　　　　　　　　　　　　　　　　adv.　　在上头,高高地
18. vertical [ˈvəːtik(ə)l]　　　　　　　　 adj.　　 垂直的,竖的
19. hollow [ˈhɔləu]　　　　　　　　　　adj.　　 空心的
　　　　　　　　　　　　　　　　　　n.　　　凹洞
　　　　　　　　　　　　　　　　　　vt.　　 挖空
20. rocker [ˈrɔkə]　　　　　　　　　　　n.　　　摇杆,摇臂
21. gravity [ˈɡræviti]　　　　　　　　　n.　　　重力,(地心)引力

22. ring [riŋ]	n.	环
	vt.; vi.	围住
23. wipe [waip]	vt.; n.	揩,擦
24. excess [ikˈses]	n.	过分,过量,剩余
	adj.	剩余的,过分的
25. recirculation [ˌriːsɜːkjʊˈleɪʃ(ə)n]	n.	再循环;重复循环

Phrases and Expressions

1. timing gear — 正时齿轮
2. oil-spray hole — 喷油孔
3. rocker-arm shaft — 摇臂轴
4. metering device — 计量装置
5. antiscuff additive — 防磨损添加剂
6. corrosion inhibitor — 防腐剂
7. detergent-dispersant — 清洁—分散剂
8. viscosity index — 黏度指数
9. foam inhibitor — 泡沫抑制剂

Notes

①although various combination of these two systems are used.
不过,这两种润滑系是以多种结合的方式使用的。
although 连接词,意为"不过"。

②... of the block = ... of the cylinder block

③... in the head = ... in the cylinder head

Exercise One

Put the following expressions into Chinese.

1. corrosion inhibitor
2. abrasive product
3. lubricating system
4. oil pump drive shaft
5. optimum conditions of lubrication
6. overhead valve engine

Exercise Two

Choose one of the four words or phrases that best matches the definition according to the text.

1. Without the aid of _____, an automobile could not move itself.
   a. friction   b. fuel   c. lubrication   d. cooling
2. The wear of two metal surfaces can be reduced to minimum by use of _____.
   a. a suitable lubricant   b. a proper coolant

  c. a sealing agent   d. a hydraulic fluid
3. The aim of the _____ is reduced metal to metal contact.
  a. cooling   b. lubrication   c. injection   d. ignition
4. An oil of _____ will flow more easily than an oil of high viscosity.
  a. low temperature   b. low pressure
  c. low viscosity   d. low tension

Exercise Three

Answer the questions:
1. Which types does the lubricating system include?
2. What are the purposes of the lubricating system?

Lesson 6

Engine Cooling System

  The burning of the fuel/air mixture inside the cylinders of a car engine produces a lot of heat. This heat must be removed to prevent the parts from expanding too much and sticking together or cracking. So the purpose of the cooling system is to keep the engine at its most efficient operating temperature at all speeds under all driving conditions.

  As fuel is burned in the engine, about one-third of the heat energy in the fuel is converted into power. Another third goes out the exhaust pipe unused, and the remaining third must be handled by the cooling system.

  The most common method is to pass water through the engine and so transfer the heat to the water. Then cold air is blown over the water to transfer the heat to the air.

  In a water-cooled engine as shown in Fig. 1-6-1 the cylinders are surrounded by a water jacket. This is a system of water passages cast into the engine block, through which water is pumped by a water pump. Automobile engine water pumps are of many designs, but most are the centrifugal type.①They consist of rotating fan or impeller, and seldom are of the positive displacement type that use gears or plungers. The hot water from the water jacket is then through the top hose into a radiator where it is cooled by the air outside before it is pumped back through the bottom hose into water jacket again.

  When an engine has just been started there is no need to cool it, so a thermostat is included to turn off the flow of water to the radiator. When the engine temperature raises 82-91℃, the thermostat starts to open, allowing fluid to flow through the radiator. By the time the coolant reaches 93℃-103℃, the thermostat is open all the way.

  The radiator is a type of heat exchanger and transfers the heats in the water to the air outside car so that the cooled water may be passed back into water jacket to extract more heat. It consists of a header tank (at the top) and a bottom tank separated by a lot of narrow tubes through which the hot water falls. Cold air passes over the tubes and the heat flows from the water to the air.

Part 1　Engineering Machinery Bases

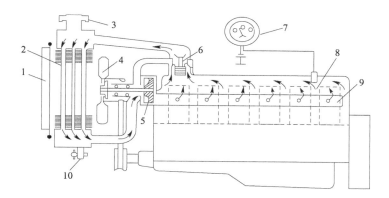

Fig. 1-6-1　Water cooling system

1-shutter;2-radiator;3-radiator cover;4-fan;5-water pump;6-thermostat;7-water temperature gauge;8-water jacket;9-connecting pipes;10-drain tap

　　The radiator is fitted at front of the car so that the car's movement provides the air flow, but when the car is stationary a fan ensures that the cooling process continues. Although this fan is sometimes driven by an electric motor, it is usually attached to the water pump and driven by the engine.

　　The water flows from the water jacket to the radiator through rubber hoses.

　　In time these are liable to perish and crack, so they must be checked regularly and replaced if they are found to be damaged.

New Words

1. remove[ri'muːv]	vt.	搬开;移动,消除
	vi.	移动,迁移
2. stick[stik]	vt.;vi.	插,粘贴,阻塞
3. crack[kræk]	vt.;vi.	破裂
	n.	裂缝
4. underestimate[ˌʌndə'estimeit]	vt.	低估
	n.	估计不足
5. centrifugal[ˌsentri'fjuːg(ə)l; sen'trifjug(ə)l]	adj.	离心的
6. impeller[im'pelə]	n.	推进者,推进器;叶轮
7. hose[həuz]	n.	软管,皮带管
8. radiator['reidieitə]	n.	散热器
9. thermostat['θəːməstæt]	n.	节温器,恒温器
10. extract['ekstrækt]	vt.	抽出,提取
11. narrow['nærəu]	adj.	窄小的
12. stationary['steiʃ(ə)n(ə)ri]	adj.	不动的
	n.	固定物

13. liable [ˈlaɪəb(ə)l]　　　　　　　　　adj.　　　易的
14. perish [ˈperiʃ]　　　　　　　　　　vt. ;vi.　　消灭,毁坏
15. regularly [ˈreɡjəlɪ]　　　　　　　　adv.　　　定期地;有规律地,经常

Phrases and Expressions

1. water-cooled engine　　　　　　　　　　　　水冷式发动机
2. be liable to　　　　　　　　　　　　　　　　易于
3. water jacket　　　　　　　　　　　　　　　 水套
4. header tank　　　　　　　　　　　　　　　 上水箱
5. bottom tank　　　　　　　　　　　　　　　 下水箱

Notes

①They consist of a rotating fan or impeller, and seldom are of the positive displacement type that use gears or plungers.

它们由一个旋转叶片,或叶轮组成,而很少用由齿轮或柱塞构成的容积式水泵。

Exercise One

Put the following expressions into Chinese.

1. cooling water circulation
2. engine cooling system
3. water cooled engine
4. air condition system
5. air cooled engine
6. crankcase ventilation

Exercise Two

Choose one of the four words or phrases that best matches the definition according to the text.

1. There are _____ types cooling systems of the engine.
   a. one　　　　　b. two　　　　　c. three　　　　　d. four
2. There are _____ types lubricating systems of the engine.
   a. one　　　　　b. two　　　　　c. three　　　　　d. four
3. Automobile engine water pump are of many designs, but most are the _____.
   a. Wheel type　　　　　　　　　　b. gear type
   c. centrifugal type　　　　　　　　d. plunger type
4. The most common cooling method is to pass the _____ through the engine and so transfer the heat.
   a. lubricating oil　　b. air　　　　c. fluild　　　　d. water

Exercise Three

Answer the questions:

1. What is the function of the cooling system?

2. What are the components that compose the engine cooling system?

Lesson 7

Engine Starting System

Ease of starting is one of the major performance characteristics of internal combustion engines. To start such an engine, it is necessary to spin the crankshaft with sufficient speed for good mixing of air and fuel and adequate compression and ignition of the combustible charge. The minimum speed with which the crankshaft of an engine should be rotated to ensure reliable starting of the engine is referred to as the cranking speed. It depends on the engine type and starting conditions.

The starting system is combination of mechanical and electrical parts that work together to start the engine. The starting is designed to change the electrical energy, which is being stored in the battery, into mechanical energy. To accomplish this conversion, a starter or cranking motor is used.

The starting speed is 40-50 rpm for carburettor engines and 150-250 rpm for diesel engines. Cranking the engine with a lower speed makes it more difficult for the engine to start, for in this case the charge has more time to escape through leaky joints and give off its heat of compression to the engine components, as a result of which both the pressure and temperature of the charge at the end of the compression stroke are reduced.

Much effort is required to crank the engine during starting, since it is necessary to overcome friction in the moving engine components and the resistance offered by the charge being compressed. The effort depends on the engine temperature, growing higher with decreasing temperature, because of the increasing viscosity of lubricating oil.

Diesel and gasoline engines are not self-starts. In order to start them, the engine crankshaft must be turned over by some outside means. Internal combustion engines may be started by the following methods:

1) hand starting;
2) electric starter motor.

Electric motor starting is the most common method used for starting automobile engine, with the development of the starting motor, the complete electric system came into existence. For example, a storage battery was then necessary to furnish the current to operate the starter motor. Also, a generator was needed to charge the battery in order to replace the electrical energy used by the starter.

The starting system provides the power to turn the internal combustion engine over until it can operate under its own power. To perform this task, the starting motor receives electrical power from the battery, and it converts this energy into mechanical energy, which transmits through the drive mechanism to the engine's flywheel.

The typical starting system has five components: battery, starting switch, starter drive, starter solenoid (relay) or switch, and starting motor (Fig. 1-7-1).

The battery serves as an energy "bank". It receives energy from the charging system and stores it until needed. Then, it supplies electrical energy in the form of current flow for the starting circuit. The starting motor is a compact but very powerful direct-current electric motor designed to crank the engine fast enough for it to start. It rotates a small gear called a pinion. A starter drive assembly connects the small pinion gear to the end of the starter motor. When the driver turns the key switch, the pinion gear meshes with the ring gear on the engine' flywheel, which drives the flywheel and cranks the engine.

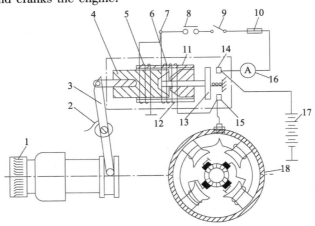

Fig. 1-7-1 Electro-magnetic starting system

1-drive pinion; 2-returning spring; 3-fork; 4-movable core; 5-maintaining solenoid; 6-attractive solenoid; 7-binding post; 8-starting button; 9-starting switch; 10-fuse; 11-copper sleeve; 12-back plate; 13-contact plate; 14,15-binding post; 16-ammeter; 17-battery; 18- direct-current electric motor

The starting system is convenient to operate and considerably eases the driver's work, but requires skilled maintenance and has only a small store of energy, which limits the number of possible starting attempts. [①]Many engines using an electric starting system are also provided with means for hand starting in case the storage battery or the starter motor should fail.

The solenoid is a magnetic switch mounted on the top of the starter motor. It has two important functions: it controls the electric circuit between the battery and starter motor and it shifts the pinion gear in and out of mesh with the ring gear.

To facilitate the starting of diesel engines under low temperature conditions, use is made of the compression release (decompressor) mechanism and heating devices.

New Words

1. spin [spin] 　　　　　　　　　　　vt. ;vi.　　　　自转;旋转
   (spun, spinning)
2. crank [kræŋk]　　　　　　　　　　 vi.　　　　　　转动曲柄,起动;开动
3. resistance [ri'zist(ə)ns]　　　　　　n.　　　　　　阻力,反抗,抵抗
4. admit [əd'mit]　　　　　　　　　　 vi.　　　　　　接纳,让……进入
5. admission [əd'miʃ(ə)n]　　　　　　n.　　　　　　接纳;承认,进气,供给
6. displacement [dis'pleism(ə)nt]　　　n.　　　　　　容量,排量

7. generator[ˈdʒenəreitə] n. 发电机,发动机,发生器
8. charge[tʃɑːdʒ] vi. 充电,充气
9. mechanical[miˈkænik(ə)l] adj. 机械的,机械制的
10. typical[ˈtipik(ə)l] adj. 典型的;代表性的
11. draw[drɔː] vi. 吸收;汲取,拖;拉
12. considerably[kənˈsid(ə)rəbli] adv. 很;颇,相当大地
13. facilitate[fəˈsiliteit] vt. 使容易;便于,促进
14. remotely[riˈməutli] adv. 遥远地;远距离地,间接地
15. pinion[ˈpinjən] n. 小齿轮,传动齿轮
16. disengage[ˌdisinˈgeidʒ] vt. 脱离,分离

Phrases and Expressions

1. be referred to as 称为,被认为是
2. turn over 转动,使转动
3. physical strength 体力
4. in respect to(of) 关于;就……而论
5. compression ratio 压缩比
6. starter solenoid 起动机电磁线圈
7. reduction gear 减速器

Notes

①Many engines using an electric starting system are also provided with means for hand starting in case the storage battery or the starter motor should fail.
许多使用电力起动系统的发动机也配有手动装置,以防蓄电池或者起动机发生故障。

Exercise One

Translate the following expressions into Chinese.
1. the diesel engine flywheel ring gear
2. starting-engine drive gear
3. electric starter motor
4. compression ratio
5. starter solenoid
6. centrifugal force of the rotating flyweight

Exercise Two

Choose one of the four words or phrases that best matches the definition according to the text
1. The escape of air and loss of the heat both result in a _____ at the end of compression.
   a. higher temperature   b. higher pressure
   c. lower temperature   d. lower compression ratio

2. The _____ must supply a great deal of current flow to the starter before it can crank the engine.
   a. generator
   b. distributor
   c. breaker
   d. battery
3. The _____ is an electromagnetically operated heavy duty switch.
   a. ignition primary coil
   b. starter solenoid
   c. ignition secondary coil
   d. breaker
4. Many solenoids also shift the cranking pinion into the _____.
   a. flywheel ring gear
   b. starter
   c. crankshaft
   d. camshaft

Exercise Three

Answer the questions:
1. What is the purpose of the starting system?
2. Which do the starting methods include?

Lesson 8

Gasoline Engine Ignition System

The ignition system is that part of the electrical system which carries the electrical current to the spark plug when the spark necessary to ignite the fuel air mixture in the combustion chamber is produced.

Many means are employed to produce the necessary high voltage required to jump a set gap. All of which are based on the principle of mutual electromagnetic induction.

Three types of ignition system are used: battery (standard, conventional) ignition system; the electronic ignition system and electronic spark advance (micro-computer control) ignition system. The battery ignition system has been used on automobile for over 60 years. Manufacturers began to use electronic ignition system on high-performance vehicles in the 1960s. However, solid-state system did not appear on any domestic passenger vehicles until early in the 1970s. These systems are now rule instead of the exception, due to stricter emission control standard and need for improving fuel economy.

The automotive ignition system has two basic functions: it must control the spark and time of the spark plug firing to match varying engine requirements, and it must increase battery voltage to a point where it will overcome the resistance offered by the spark plug gap and fire the plug.

Modern ignition systems operate from a battery. Conventional system consists of the battery, ignition coil, distributor, condenser, ignition switch, spark plug, resistor and the necessary low and high tension wiring.

Fig. 1-8-1 is shown a typical battery ignition system for a six-cylinder engine.

An automobile ignition system is divided into two electrical circuits—the primary and secondary circuits. The primary circuit carries low voltage. This circuit operates only on battery current and is controlled by the breaker points and ignition switch.

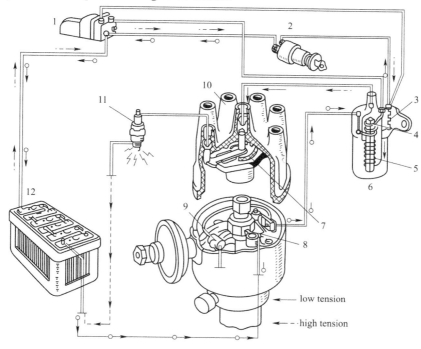

Fig. 1-8-1 Battery ignition system

1-starter switch; 2-ignition switch; 3-series winding; 4-primary winding; 5-secondary winding; 6-ignition coil; 7-distributor; 8-breaker; 9-condenser; 10-distributor cap; 11-spark plug; 12-battery

The secondary circuit consists of the secondary winding in the coil, the high tension lead between the distributor and coil, the distributor cap, distributor rotor, the spark plug leads and the spark plug.

The primary ignition circuit starts at the battery and passes through the ammeter, the ignition switch, the heavy or "primary" winding about the soft-iron laminated core of the coil, through the ignition points to ground. One end of the condenser is attached to the primary circuit and the other end is grounded.

The secondary coil winding is not connected electrically to the primary. It starts from ground in the coil, runs through about 21000 turns of fine wire, and then passes through a heavily insulated wire into the center of distributer cap. A carbon contact carries the current to the rotor, which, as it revolves, distributes the current to the six segments which, in turn, send it to the spark plugs through the spark plug wires. After the current jumps the spark plug, igniting the gasoline mixture, it is grounded.

The distributor shaft on which the rotor is placed is turned by the engine camshaft. At the top of this shaft, just under the rotor,①there is a six-lobe cam, one for each of the six cylinders in the engine shown in Fig. 1-8-1 (The number of lobes on the cam is equal to the number of cylinders up to eight. When a single set of ignition points is used, twelve- and sixteen-cylinder engines use two sets of points in combination with six-and eight-lobe cams).

In the early 1970s many American vehicles began to install solid-state electronic ignition system. This action is necessary not only to meet stricter emission control standard but also increase the fuel economy of the vehicle. The electronic ignition system does not use the contact points as a switching device to open and close the primary circuit. Instead, the system uses an electronic switch in the form of one or more transistors to control primary current flow. As a result, the system can provide overall higher voltages necessary to fire the leaner air/fuel mixture on emission controlled engine and also bridge enlarged air gaps due to plug wear. Consquently, advantages of the electronic ignition system include:

1) Greater available secondary voltage, especially at high engine.
2) Reliable and consistent system performance at any and all engine speeds.
3) A potential for more responsive and variable ignition advance curve.
4) Decreased maintenance cost of the system.

Electronic spark advance ignition system has incorporated computer technology and developed from single-control system to modern centralized control system. It includes various sensors, electronic control unit (ECU) and ignition actuator. It can precisely control ignition timing, energizing period and anti-detonation.

New Words

1. current [ˈkʌr(ə)nt] n. 流;电流
2. gap [gæp] n. 间隙;火花隙
3. mutual [ˈmjuːtʃuəl] adj. 相互的;共同的
4. electromagnetic [iˌlektrə(u)mægˈnetik] adj. 电磁的
5. induction [inˈdʌkʃ(ə)n] n. 吸入;引入,感应
6. battery [ˈbætri] n. 电池;蓄电池
7. coil [kɔil] n. 线圈,线组
8. primary [ˈpraim(ə)ri] adj. 第一位的;最初的,基本的
   n. 原线圈;初级线圈,底色
9. particular [pəˈtikjulə(r)] adj. 特殊的,个别的
   n. 特点
10. ammeter [ˈæmitə] n. 电流计;安培计
11. winding [ˈwaindiŋ] n. 绕组
12. laminated [ˈlæmineitid] adj. 叠层的,叠片的
13. core [kɔː] n. 铁芯
14. condenser [kənˈdensə] n. 电容器,冷凝器
15. ground [graund] n. 地,土地
    vt.;vi. 接地;落地
16. secondary [ˈsek(ə)nd(ə)ri] adj. 第二的,次级的

Part 1 Engineering Machinery Bases

	n.	副手
17. turn[tə:n]	vt.;vi.	圈,匝;转动
	n.	转动,转变
18. insulate['insjuleit]	vt.	隔绝
19. distributor[di'stribjutə]	n.	分电器;配电器
20. contact['kɔntækt]	n.	接触;触点,接触器
21. distribute[di'stribju:t]	vt.	分配,配电
22. rotor['rəutə]	n.	转子,转片,分火头
23. segment['segm(ə)nt]	n.	整流子片;侧电极
24. thence[ðens]	adv.	从这里起,从那时起,因此
25. camshaft['kæmʃɑ:ft]	n.	凸轮轴
26. lobe[ləub]	n.	凸角;凸棱
27. cam[kæm]	n.	凸轮
28. transistor[træn'zistə]	n.	晶体管

Phrases and Expressions

1. battery-ignition system　　　　　　蓄电池点火系
2. electronic-ignition system　　　　　电子点火系
3. electronic spark advance ignition system(ESA)　　电子点火提前控制系统(微机控制点火系统)
4. electronic control unit(ECU)　　　电子控制单元
5. solid-state system　　　　　　　　固态系统(晶体管系统)
6. soft-iron laminated core　　　　　　软铁叠片铁芯
7. six-lobe cam　　　　　　　　　　六棱凸轮
8. in turn　　　　　　　　　　　　　依次
9. be equal to　　　　　　　　　　　等于……;和……相等

Notes

①...,there is a six-lobe cam,one for each of the six cylinders in the engine shown in Fig.1-8-1 shown in Fig.1-8-1 是分词短语作定语,说明前面的 six-lobe cam。

Exercise One

Put the following expressions into Chinese.

1. the combustible mixture
2. electronic-ignition system
3. mutual induction
4. the electrodes of the spark plug
5. a primary circuit and a secondary circuit
6. spark plug

Exercise Two

Choose one of the four words or phrases that best matches the definition according to the text

1. The high-voltage surges necessary to produce the sparks at the spark plug are provided by the _____ components.
   a. starting system   b. ignition system
   c. lubricating system   d. cooling system

2. The compressed air-fuel mixture in the engine cylinder is ignited by the _____.
   a. battery   b. generator
   c. spark   d. solenoid

3. The magnetic saturation of the secondary coil depends upon the _____ applied to the primary coil.
   a. amount of electrical current   b. direction of electrical current
   c. vibration of electrical current   d. frequency of electrical current

4. The magnetic field produced by the primary coil winding must completely saturate the _____.
   a. ignition coil   b. secondary winding
   c. spark plug   d. battery

Exercise Three

Answer the questions:

1. Which kind of the ignition does the gasoline engine use?
2. What is the purpose of the ignition system?

Lesson 9

Clutches

The engine produces the power to drive the vehicle. The drive line transfers the power of the engine to the wheels. The drive line consists of the parts from the back of the flywheel to the wheels. These parts include the clutch, the transmission, the drive shaft, and the final drive assembly.

A clutch is a friction device used to connect and disconnect a driving force from a driven member. In automotive applications, it is used in conjunction with an engine flywheel to provide smooth engagement and disengagement of the engine and manual transmission.

There are a number of clutches used in different types of the vehicles. They are classified as under:

1. Friction clutch:
1) single plate clutch;
2) multiplate clutch.
(1) wet;
(2) dry.

2. Coil pressure spring clutch;

3. Diaphragm clutch;

4. Positive clutch.

1) dog;

2) spline type.

Since a internal combustion engine develops little power or torque at low rpm, it must gain speed before it will move the vehicle. However, if a rapidly rotating engine is suddenly connected to the drive line of a stationary vehicle, a violent shock will result.

①So gradual application of load, along with some slowing of engine speed, is needed to provide reasonable and comfortable starts. In vehicles equipped with a manual transmission, this is accomplished by means of a mechanical clutch.

The clutch utilizes friction for its operation. The main parts of the clutch are a pressure plate and a driven disk. The pressure plate is coupled with flywheel, while the driven disk is fitted to the transmission input shaft. The pressure plate is pressed to the disc by the springs so that the torque is transmitted owing to friction forces from the engine to the input shaft of the transmission.

The clutch has two members positively driven by the engine and a third attached to the transmission shaft. When these members are separated, by pushing down the clutch pedal, the engine will run without turning the transmission shaft, thus permitting the gears to be shifted easily, ②the engine to idle, or the car to be stopped without stalling the engine. The friction surfaces of the clutch are designed so that the driven member slips on the others when the pressure is first applied. As the pressure is increased, the driven member is brought gradually to the speed of the driving members. When the speed of the three members becomes equal, slippage ceases entirely, ③the three making firm contact.

The drive is accomplished by the friction between the three members, which depends upon the material in contact and the pressure forcing them together. This force is maintained by spring pressure and must be sufficient to prevent slipping when the clutch is engaged fully, and the surfaces must be of such material as to provide sufficient friction to carry the load. When the clutch pedal is pushed down, the spring or springs are compressed, thus freeing the engine from the transmission line.

Fig. 1-9-1 shows a typical clutch that has been taken apart. The principal parts are the driving members, the driven member and the operating members.

One driving member consists of a cover which carries a cast-iron pressure plate or driving disc, the pressure springs, and the releasing levers. The entire assembly is bolted to the flywheel and rotates with it at all times. The flywheel acts as a second driving member, the flywheel and the pressure plates gripping the driven member between them under the action of the pressure springs. To dissipate the heat generated by friction properly in operation of the clutch, the clutch housing or cover is provided with openings for ventilation.

The driven member consists of a disk or plate which is free to slide lengthwise on the splines of the clutch shaft④but which drives the shaft through those same splines. The clutch disc carries friction material on both bearing surfaces.

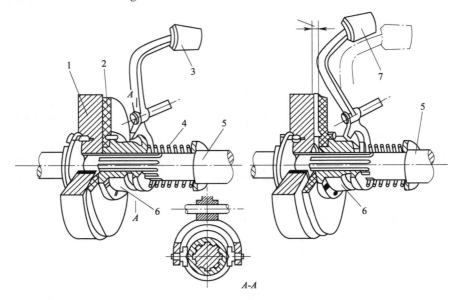

Fig. 1-9-1 Operation of friction clutch

1-flywheel;2-driven plate;3,7-pedal;4-spring;5-driven shaft;6-driven hub

The operating mechanism consists of the foot pedal, the linkage, the release or throw out bearing, the release levers and the springs necessary to insure the proper operation of the clutch.

New Words

1. clutch[klʌtʃ]	n.	离合器
	vt.;vi.	抓住
2. conjunction[kənˈdʒʌŋ(k)ʃ(ə)n]	n.	连接,连词(语法)
3. disengagement[disinˈgeidʒm(ə)nt]	n.	断开,脱离,释放
4. manual[ˈmænju(ə)l]	adj.	用手的,手动的
5. violent[ˈvaiəl(ə)nt]	adj.	猛烈的,激烈的
6. positively[ˈpɔzətivli]	adv.	刚性地,强制地
7. transmission[trænzˈmiʃ(ə)n]	n.	传动,传动装置,变速器
8. shift[ʃift]	vt.;vi.;n.	改变,换挡,位置
9. idle[aidl]	adj.	空转的;惰的,闲置的
10. stall[stɔːl]	vt.;vi.	空转
11. friction[ˈfrikʃ(ə)n]	n.	摩擦
12. slip[slip]	vt.;vi.	滑倒;滑掉
	n.	滑倒;滑动
13. slippage[ˈslipidʒ]	n.	滑动,打滑;空转
14. cease[siːs]	vt.;vi.;n.	停止,中止

Part 1 Engineering Machinery Bases

15.	entire [inˈtaiə]	adj.	完全的,全部的
		n.	全体
16.	maintain [meinˈtein]	vt.	维护;维持,保持
17.	spring [spriŋ]	n.	弹簧;弹力,跳跃
	(sprang, sprung)	vt.; vi.	弹起,跳跃
18.	sufficient [səˈfiʃ(ə)nt]	adj.	充分的,足够的
19.	exertion [igˈzəːʃn]	n.	费力;尽力,发挥
20.	exclusively [iksˈkluːsivli]	adv.	专门地,排他地
21.	cast-iron [kɑːstˈaiən]	n.	铸铁
		adj.	铸铁的;铸铁制的,硬的,刚毅的
22.	disc [disk]	n.	片;盘
	(disk)		
23.	release [riˈliːs]	vt.; n.	释放,放松
24.	lever [ˈliːvə]	n.	杆;手柄
25.	grip [grip]	vt.; vi.	夹住;夹紧
		n.	夹住;夹具
26.	dissipate [ˈdisipeit]	vt.; vi.	消除;消散
27.	housing [hauziŋ]	n.	套;壳
28.	opening [ˈəup(ə)niŋ]	n.	窗孔,开口
29.	ventilation [ˌventiˈleiʃ(ə)n]	n.	换气,通风
30.	slide [slaid]	vt.; vi.	滑动;滑移
	(slid, slide/slidden)		
31.	lengthwise [ˈleŋθwaiz]	adv.	纵向地;沿长度方向
32.	spline [splain]	vt.	开槽;用花键连接;嵌键
		n.	花键;方栓;齿槽
33.	linkage [ˈliŋkidʒ]	n.	杠杆传动装置;联动装置;杆系
34.	insure [inˈʃɔː; inˈʃuə]	vt.	保险;保证

Phrases and Expressions

1. in conjunction with　　　　　　　　和……一起;连同……一起
2. along with　　　　　　　　　　　　与……一起;除……之外
3. single plate clutch　　　　　　　　 单片离合器
4. multiplate clutch　　　　　　　　　多片离合器
5. coil pressure spring clutch　　　　螺旋弹簧式离合器
6. diaphragm clutch　　　　　　　　　膜片式离合器
7. positive clutch　　　　　　　　　　刚性连接式离合器
8. driving member　　　　　　　　　　主动件

9. driven member 被动件
10. release bearing 分离轴承
11. clutch housing 离合器壳
12. throw out 分离,断开,投出
13. act as 作为;充当;起……作用

Notes

①So gradual application of load, along with some slowing of engine speed, is needed to provide reasonable and comfortable starts.

随着负荷的增加,发动机的转速也会降低。因此,需要提供合理平稳的起动。

②...,the engine to idle,... 发动机空转

the engine 前省略 permitting,避免重复。the engine 是 permitting 的宾语,to idle 是宾语补足语。

③...the three making firm contact. ……三部件紧紧接触。

the three 是分词 making 的逻辑主语,此分词短语是分词独立结构,作结果状语。

④...but which drives...

此处 which 指 a disc or plate,引出定语从句。

此处 but 是从属连接词,意为"若不……","如果不……"。

Exercise One

Put the following expressions into Chinese.

1. maximum torque of the engine
2. power developed by the engine
3. clutch maintenance service
4. clutch pressure spring
5. hydraulic clutch
6. clutch pedal clearance

Exercise Two

Choose one of the four words or phrases that best matches the definition according to the text.

1. To drive the wheel of the automobile the power developed by _____ must be transmitted to them.

    a. battery   b. generator

    c. engine   d. starter

2. A clutch is used to disengage the driven member from engine to the _____ as the required gear is being selected.

    a. brake   b. steering wheel

    c. steerable wheel   d. gearbox

3. The clutch is usually attached directly to the engine _____.

    a. crankshaft   b. flywheel

c. camshaft   d. drive shaft

4. The spring pressure must be sufficient to prevent _____ when the clutch is fully engaged.

    a. applying more force   b. vibration

    c. slippage   d. joint

Exercise Three

Answer the questions:

1. What is the purpose of the clutch?
2. What components does the drive line include?

Lesson 10

Transmissions

A transmission is a speed and power changing device installed at some point between the engine and driving wheels of the vehicle. It provides a means for changing the ratio between engine rpm (revolutions per minute) and driving wheels rpm to best meet each particular driving situation.

The transmission is designed for changing the torque transmitted from the engine crankshaft to the propeller shaft, reversing the vehicle movement and disengaging the engine from the drive line for a long time at parking or coasting. A higher torque should be applied to the wheels to set an automobile in motion or move uphill with full load than to keep it rolling after it gets under way on level stretches of the road, when inertia is high and tractive resistance is low. To meet these variable torque requirement, special gear boxes are used. Such gear boxes are called transmissions.

According to gear ratios changing, transmissions can be divided into three types.

1. fixed-ratio transmission: having gear drives and some fixed-ratios.

1) fixed-shaft gear type;

2) planetary gear type.

2. continuously variable speed transmission: having continuous, variable ratios.

1) hydraulic-control type (drive part: torque converter);

2) electric-control type (drive part: direct current series motor).

3. combination type transmission: hydraulic-mechanical type formed by torque converter and fixed-shaft gearbox.

Fig. 1-10-1 shows sliding-mesh 3-speed gearbox.

The gear box consists of a housing, an input shaft and gear, an output and gear, an idler shaft, a reverse gear, a cluster of gears and a gear shift mechanism.

The engine torque is conveyed from the driven disc of the clutch by the clutch (primary or first motion) shaft. The clutch shaft revolves in a bearing in the gearbox casing and has an integral pinion which is in permanent engagement with a corresponding pinion on the layshaft below it,[①] the two being termed the constant mesh pinions.

The layshaft consists of a cluster of four pinions, including the constant-mesh pinion, rotating together either upon a fixed layshaft spindle or in bearings in the gearbox casing.

The splined mainshaft is carried in a spigot bearing in the clutch shaft and in a bearing in the gearbox casing at the rear end, and is coupled to the propeller shaft through a universal joint. Sliding on the mainshaft are two splined pinions having integral collars into which fit the selector forks.

Top Gear or Direct Drive

The second-gear pinion has projecting "dogs" on the side facing the clutch shaft, and[②] when slid along the splined mainshaft by the selector fork, the dogs mate with corresponding projections on the clutch shaft to give a positive "direct" drive between the two shafts for top gear (Fig. 1-10-1d).

Second Gear

For second gear the mainshaft pinion is slid into engagement with the second-gear pinion on the layshaft. The drive is "indirect" and passes from the clutch shaft through the constant-mesh gears to the layshaft and back to the mainshaft through the second-gear pinions (Fig. 1-10-1c).

The gear ratio is the product of the gear ratio of the two pairs of pinions. thus:

$$\text{Gearbox ratio on 2nd gear} = \frac{\text{Rev/s of clutch shaft}}{\text{Rev/s of mainshaft}}$$

$$= \frac{\text{Rev/s of clutch shaft}}{\text{Rev/s of layshaft}} \times \frac{\text{Rev/s of layshaft}}{\text{Rev/s of mainshaft}}$$

$$= \frac{{}^{③}\text{No. teeth on layshaft c.m. pinion}}{\text{No. teeth on clutch shaft c.m. pinion}} \times \frac{\text{No. teeth on mainshaft 2nd gear}}{\text{No. teeth on layshaft 2nd gear}}$$

This is more easily remembered by taking the pinions in pairs in their order from the mainshaft to the clutch shaft. Thus:

$$\text{Gearbox ratio} = \frac{\text{Teeth on mainshaft pinion}}{\text{Teeth on corresponding layshaft pinion}} \times \frac{\text{Teeth on layshaft c.m. pinion}}{\text{Teeth on clutch shaft c.m. pinion}}$$

First Gear

In this case the first-gear pinion on the mainshaft is engaged with the corresponding first-gear pinion on the layshaft and due to their respective sizes a greater gear reduction is obtained than with the second-gear pinions. The ratio is calculated in a similar manner, using the number of teeth on the first-gear pinions (Fig. 1-10-1b).

Reverse Gear

The fourth pinion on the layshaft is permanently engaged with a reverse idler pinion turning on a short shaft fixed in the gearbox casing. When the first-gear mainshaft pinion is moved into engagement with this idler pinion the direction of rotation of the mainshaft is reversed. In calculating the reverse-gear ratio, the number of teeth on the first-gear mainshaft pinion and on the layshaft reverse-gear pinion are taken, since the size of the idler pinion is immaterial—to the gear ratio (Fig. 1-10-1e).

Part 1 Engineering Machinery Bases

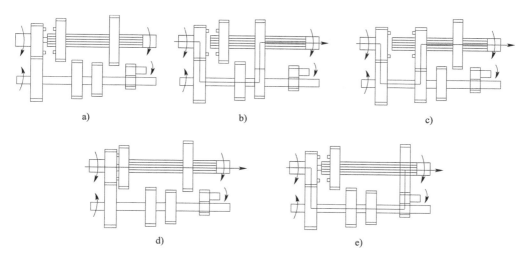

Fig. 1-10-1 Sliding-mesh 3-speed gearbox
a) neutral; b) first gear; c) second gear; d) third gear; e) reverse gear

Automatic gearbox is commonly used in modern car and engineering machinery. It made driving much easier. It has three basic systems—a torque converter, a gear system, and a hydraulic system.

These fit together in a unit that fastens directly behind the engine. The torque converter is like the clutch in a manual gearbox.

The hydraulic system is the "brain" of the gearbox. This section must know when to change from one gear to another. It must also provide the fluid pressure needed to apply or release the parts of gear section so that the gears shift at the correct time.

The hydraulic section is a complex maze of passages and valves that senses speed and load.

The gear section has gears that change the gear ratios for torque multiplication and gear reduction. This is accomplished through the use of planetary gear sets. It also has the input and output shaft needed to transmit power to the drive wheels. The computer uses sensors on the engine and transmission to detect such things as throttle position, vehicle speed, engine speed, engine load, stop light switch position, etc. to control exact shift points as well as how soft or form the shift should be.

Another advantage is that they have a self diagnostic mode which can detect a problem early on and worn you with an indicator light on the dash. A technician can then plug test equipment in and retrieve a list of trouble codes that will help pinpoint where the problem is.

New Words

1. situation [sitju'eiʃ(ə)n]　　　　　　　n.　　　形势；环境；情况
2. coasting ['kəustiŋ]　　　　　　　　　n.　　　滑行；滑行距离，惯性运转
3. uphill [ʌp'hil]　　　　　　　　　　　adv.　　上坡；上坡路
　　　　　　　　　　　　　　　　　　　adj.　　上坡的，艰难的
4. stretch [stretʃ]　　　　　　　　　　 n.　　　一段路程，距离，路段

5. inertia [iˈnəːʃə]　　　　　　　　n.　　　　　惯性,迟钝
6. casing [ˈkeisiŋ]　　　　　　　　n.　　　　　壳;套
7. gearbox [ˈgiəbɔks]　　　　　　　n.　　　　　变速器;齿轮箱
8. aluminium [æl(j)uˈminiəm]　　　n.　　　　　铝
9. malleable [ˈmæliəb(ə)l]　　　　 adj.　　　　可锻的,韧性的
10. alternatively [ɔːlˈtəːnətivli]　adv.　　　　要不;两者挑一
11. crankcase [ˈkræŋkkeis]　　　　 n.　　　　　曲轴箱
12. convey [kənˈvei]　　　　　　　 vt.　　　　　输送,递交,传递
13. permanent [ˈpəːm(ə)nənt]　　　 adj.　　　　永久的;不变的,不动的
14. engagement [inˈgeidʒm(ə)nt]　　n.　　　　　啮合
15. corresponding [ˌkɔriˈspɔndiŋ]　adj.　　　　相应的
16. layshaft [ˈleiʃæft]　　　　　　 n.　　　　　中间轴,副轴
17. term [təːm]　　　　　　　　　　vt.　　　　　把……叫作
　　　　　　　　　　　　　　　　　　n.　　　　　学期,术语,期限
18. constant [ˈkɔnst(ə)nt]　　　　 adj.　　　　不变的,恒定的
19. cluster [ˈklʌstə]　　　　　　　n.　　　　　组;群
20. mainshaft [meinʃɑːft]　　　　　n.　　　　　主轴;第二轴;输出轴
21. spigot [ˈspigət]　　　　　　　 n.　　　　　栓;插销,连接器
22. couple [ˈkʌp(ə)l]　　　　　　　vi.　　　　　结合;配合
23. universal [juːniˈvəːs(ə)l]　　 adj.　　　　万能的;通用的
24. collar [ˈkɔlə]　　　　　　　　 n.　　　　　环;安装环
25. selector [siˈlektə]　　　　　　n.　　　　　选择器
26. projecting [prəuˈdʒektiŋ]　　　adj.　　　　凸出的;突出的
27. mate [meit]　　　　　　　　　　vt.;vi.;n.　　配合
28. respective [riˈspektiv]　　　　adj.　　　　各自的;各个的
29. manner [ˈmænə]　　　　　　　　 n.　　　　　方法
30. idler [ˈaidlə]　　　　　　　　 n.　　　　　惰轮;空转轮
31. immaterial [iməˈtiəriəl]　　　 adj.　　　　不重要的,非物质的

Phrases and Expressions

1. drive line　　　　　　　　　　　　　　　　　　动力传动系统;驱动系统
2. under way　　　　　　　　　　　　　　　　　　行进中
3. tractive resistance　　　　　　　　　　　　　牵引阻力
4. step speed transmission　　　　　　　　　　　有级变速器
5. continuously variable speed transmission　　无级变速器
6. torque converter　　　　　　　　　　　　　　 液力变速器
7. sliding-mesh 3-speed gearbox　　　　　　　　 滑动啮合式三挡变速器
8. gearbox casing　　　　　　　　　　　　　　　 变速器壳

Part 1　Engineering Machinery Bases

9. constant-mesh pinion　　　　　　　常啮齿轮
10. splined mainshaft　　　　　　　　花键轴
11. spigot bearing　　　　　　　　　　导向轴承
12. universal joint　　　　　　　　　　万向节
13. selector fork　　　　　　　　　　　换挡拨叉
14. projecting dog　　　　　　　　　　凸牙
15. top gear　　　　　　　　　　　　　最高挡
16. second gear　　　　　　　　　　　第二挡
17. first gear　　　　　　　　　　　　第一挡
18. reverse gear　　　　　　　　　　　倒挡
19. direct drive　　　　　　　　　　　直接挡

Notes

①..., the two being termed the constant-mesh pinions.

……，这两个齿轮叫常啮齿轮。

the two 是 being termed（现在分词被动形式）的逻辑主语，整个短语是分词独立结构，作句子的附加说明。

②... when slid along the splined mainshaft by the selector fork,...

……当换挡拨叉拨动二挡齿轮，使它在花键上滑动时，……

when 和 slide 之间省略 it is, it 即为主句的主语。

③No. teeth on layshaft c. m. pinion/No. teeth on clutch shaft c. m. pinion

式中 c. m. 即为 constant-mesh。

Exercise One

Put the following expressions into Chinese:

1. automatic transmission
2. top gear
3. planetary gear set
4. overrunning clutch
5. reverse gear
6. self diagnostic mode

Exercise Two

Choose one of the four words or phrases that best matches the definition according to the text.

1. The _____ is a speed and power changing device installed at some point between the engine and driving wheels of the vehicle.

　　a. clutch　　　　　　　　　　　　b. universal joint
　　c. transmission　　　　　　　　　d. drive shaft

2. The purpose of _____ is designed to change the directions of the car.

　　a. top gear　　　　　　　　　　　b. reverse gear

      c. first gear   d. second gear

3. The _____ are most efficient and can provide more torque and power than traditional transmissions.

      a. continuously variable transmission   b. manual transmission

      c. automatic transmission   d. planetary type transmission

4. The _____ is a connection for transmitting power from a driving to a driven shaft through an angle.

      a. clutch   b. differential

      c. universal joint   d. release mechanism

Exercise Three

Answer the questions:

1. What are the purposes of the transmission?

2. Which components does the automatic transmission include?

Lesson 11

Torque Converters

Hydraulic equipment is frequently met in road-making, hoisting and transporting devices.

①Just like manual transmission, machines with automatic transmissions need a way to let the engine turn while the wheels and gears in the transmission come to stop. Manual transmission machines use a clutch, which completely disconnects the engine from the transmission. Automatic transmission machines use torque converter.

A torque converter is a type of fluid coupling, which allows the engine to spin somewhat independently of transmission. If the engine is turning slowly, such as when the vehicle is idling at a stoplight, the amount of torque passed through the torque converter is very small, so keeping the vehicle still requires only light pressure on the brake peddle.

If you were to step on the gas peddle while the vehicle is stopped, you would have to press harder on the brake to keep the vehicle from moving. This is because when you step on the gas, the engine speeds up and pumps more fluid into the torque converter, causing more torque to be transmitted to the wheels.

There are four components inside the very strong housing of the torque converter:

(1) pump;

(2) turbine;

(3) stator;

(4) transmission fluid.

The housing of the torque converter is bolted to the flywheel of the engine, so it turns at whatever speed the engine is running at. The fins that make up the pump of the torque converter are attached to the housing, so they also turn at the same speed as the engine.

Fig. 1-11-1 shows the parts of the torque converter. The pump inside a torque converter is a type of centrifugal pump. As it spins, fluid is flung to the outside. As fluid is flung to the outside, a vacuum is created that draws more fluid in at the center.

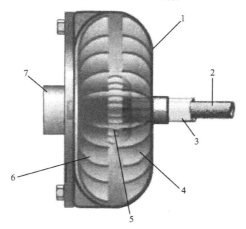

Fig. 1-11-1 The parts of a torque converter

1-torque converter housing (connects to flywheel); 2-turbine output shaft (connects to transmission); 3-stator output shaft (connects to fixed shaft in transmission); 4-pump (fixed to housing); 5-stator; 6-turbine; 7-flywheel (connects to engine)

The fluid then enters the blades of the turbine, which is connected to the transmission. The turbine causes the transmission to spin. The blades of the turbine are curved. This means that the fluid, which enters the turbine from the outside, has to change direction before it exits the center of turbine. It is this directional change that causes the turbine to spin.

The fluid exits the turbine at the center, moving in the a different direction than when it entered. The fluid exits the turbine moving opposite the direction that the pump (and engine) are turning. If the fluid were allowed to hit the pump, it would slow the engine down, wasting power. This is why a torque converter has a stator.

The stator resides in the very center of the torque converter. Its job is to redirect the fluid returning from the turbine before it hits the pump again. This dramatically increases the efficiency of the torque converter.

The stator has a very aggressive blade design that almost completely reverses the direction of the fluid. A one-way clutch (inside the stator) connects the stator to a fixed shaft in the transmission. Because of this arrangement, the stator can not spin with the fluid, it can spin only in the opposite direction, forcing the fluid to change direction as it hits the stator blades.

Something a little bit tricky happens when the vehicle gets moving. There is a point, around 40 mph (64kph), at which both the pump and the turbine are spinning at almost the same speed (the pump always spins slightly faster). At this point, the fluid returns from the turbine, entering the pump already moving in the same direction as the pump, so the stator is not needed.

Even though the turbine changes the direction of the fluid and flings it out the back, the fluid still ends up moving in the direction that the turbine is spinning because the turbine is spinning faster in one direction than the fluid is being pumped in the other direction. If you were standing in the back of a pickup moving at 60 mph, and you threw a ball out back of that pickup at 40mph, the ball would still be going forward at 20 mph. This is similar to what happens in the turbine. The fluid is being flung out the back in one direction, but not as fast as it was going to start with in the other direction.

At these speeds, the fluid actually strikes the back sides of the stator blades, causing the stator to freewheel on its one-way clutch so it doesn't hinder the fluid moving through it.

In addition to the very important job of allowing the vehicle come to a complete to stop without stalling the engine, the torque converter actually gives it more torque you accelerate out of a stop. Modern torque converters can multiply the torque of the engine by two to there times. This effect only happens when the engine is turning much faster than the transmission.

At high speeds, the transmission catches up the engine, eventually moving at almost the same speed. Ideally, the transmission would move at exactly the same speed as the engine, because this difference in speed waste power. This is part of the reason why the vehicles with automatic transmissions get worse gas mileage than that with manual transmissions.

To counter this effect, some vehicles have a torque converter with a lockup clutch. When the two halves of the torque converter get up to speed, this clutch locks them together, eliminating the slippage and improving efficiency.

New Words

1. turbine['tə:bain;-in]	n.	涡轮机
2. stator['steitə]	n.	导轮;(电机、汽轮机的)定子
3. redirect[ˌri:də'rekt]	vt.	使改变方向
4. tricky['triki]	adj.	复杂的,棘手的
5. hinder['hində]	vt.	阻碍;妨碍
6. counter['kauntə]	vt.	抵消,消除
7. aggressive[ə'gresiv]	adj.	有力的,自信的
8. eventually[i'ventʃuəli]	adv.	最终发生地
9. eliminate[i'limineit]	vt.	消除,删除
10. multiply['mʌltiplai]	v.	增加,乘

Phases and Expressions

1. torque converter   液力变矩器
2. fluid coupling   液压耦合器
3. gas peddle   加速踏板
4. one-way clutch   单向离合器
5. gas mileage   燃油里程(经济性指标)
6. get up to   达到
7. in addition to   除……之外
8. end up   最终(会)

Part 1　Engineering Machinery Bases

Notes

①Just like manual transmission, machines with automatic transmissions need a way to let the engine turn while the wheels and gears in the transmission come to a stop.

就像装有手动变速器的汽车一样,装有自动变速器的汽车需要安装一个装置,该装置使发动机在车轮和变速器齿轮停止转动时能继续运转。

Exercise One

Put the following expressions into Chinese:

1. hydraulic equipment being used in road making, hoisting and transporting devices
2. driving wheel turning the working fluid at a high speed
3. fluid coupling and torque converter
4. hydraulic clutch used to transmit engine torque to transmission gears
5. the engine loads
6. torque converter in connection with automatic transmission

Exercise Two

Choose one of the four words or phrases that best matches the definition according to the text.

1. The housing of the _____ is bolted to the flywheel of the engine, so it turns at whatever speed the engine is running at.
    a. fluid coupling   b. torque converter
    c. brake   c. transmission

2. For eliminating slippage and improving efficiency, some vehicles have a torque converter with a _____.
    a. one-way clutch   b. double-action clutch
    c. lock-up clutch   d. manual clutch

3. The _____ has a very aggressive blade design that almost completely reverses the direction of the fluid.
    a. turbine   b. pump
    c. stator   d. fluid coupling

4. A torque converter is a type of _____, which allows the engine to spin somewhat independently of transmission
    a. turbine   b. pump
    c. stator   d. fluid coupling

Exercise Three

Answer the questions:

1. Which components does the torque converter include?
2. What can the stator increase?

Lesson 12

Differentials

One important unit in the power train is the differential (Fig. 1-12-1), which is driven by the final drive. The differential is located between the axles and permits one axle to turn at different speed from that of the other. The variations in axle speed are necessary when a vehicle rounds a corner or travels over uneven ground. At the same time, the differential transmits engine torque to the drive axles. The drive axles are on a rotational axis that is 90° different from the rotational axis of the drive shaft.

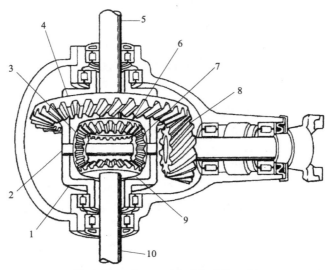

Fig. 1-12-1 Construction of the differential

1-differential case; 2-pinion shaft; 3-differential pinion; 4-ring gear; 5-axle shaft; 6-side gear; 7-differential pinion; 8-drive pinion; 9-side gear; 10-axle shaft

A differential assembly uses drive shaft rotation to transfer power to the axle shafts. The differential must be capable of providing torque to both axles, even when they are turning at different speeds. The differential assembly is constructed from the following: the differential case, the drive pinion, the ring gear, the differential gears (side gears) and the differential pinions (planetary gears, pinion gears).

The ring gear bolts to the flange on one side of the case. When the pinion drives the ring gear the differential case also turns. The case has circular holes through which the axle shafts fit. On the end of each axle shaft, there is a bevel gear called a side gear or differential gear. In the fact, the side gears are splined to the axle shafts. Whenever the side gears turn, the axle shafts turn.

There is a space between the two side gears. A shaft that is perpendicular to the axle shafts fits in this space. This is called the pinion shaft. On each end of the pinion shaft, there is a bevel gear that meshes with both side gears. We call these gears the differential pinions.

Part 1 Engineering Machinery Bases

The rear wheels of a vehicle do not always turn at the same speed. When the vehicle is turning or when tire diameters differ slightly, the rear wheels must rotate at different speeds. If there were a solid connection between each axle and the differential case, the tires would tend to slide, squeal, and wear whenever the operator turned the steering wheel of the vehicle. A differential is designed to prevent this problem.

When a vehicle is driving straight ahead, the ring gear, the differential case, the differential pinion (planetary) gears, and the differential side gears turn as a unit. The two differential pinion gears do not rotate on the pinion shaft, because they exert equal force on the side gears. As a result, the side gears turn at the same speed as the ring gear, causing both rear wheels to turn at the same speed.

When the vehicle begins to round a corner, the differential pinion gears rotate on the pinion shaft. This occurs because the pinion gears must walk around the slower turning differential side gear. Therefore, the pinion gears carry additional motion to the faster turning outer wheel on the turn.

New Words

1. differential [ˈdifərənʃəl]	n.	差别,微分,差速器
	adj.	差别的,微分的,差速的
2. axis [ˈæksis]	n.	轴,轴线,坐标轴
3. assembly [əˈsembli]	n.	装配,总成
4. flange [flrændʒ]	n.	凸缘,法兰盘
5. diameter [daiˈæmitə]	n.	直径
6. bevel [ˈbevl]	n.	斜边,斜面
	vi.	把磨成(切成)……斜角
7. planetary [ˈplænəteri]	adj.	行星的;[机] 行星齿轮的

Phrases and Expressions

1. differential assembly	差速器总成
2. side gear	差速齿轮,半轴齿轮
3. differential gear	差速齿轮,半轴齿轮
4. pinion gear	行星齿轮
5. differential pinion	行星齿轮
6. planetary gear	行星齿轮
7. ring gear	环齿(主减速器被动轮)
8. drive pinion	主减速器主动轮
9. pinion shaft	十字轴,行星齿轮轴
10. bevel gear	斜齿轮

Exercise One

Put the following expressions into Chinese:

1. final drive assembly
2. differential assembly
3. side gear
4. differential gear
5. pinion gear
6. differential pinion

Exercise Two

Choose one of the four words or phrases that best matches the definition according to the text.

1. A shaft that is perpendicular to the axle shafts fits in the space between the two side gears and is called the _____.

    a. pinion shaft　　　　　　　b. ring gear

    c. side gear　　　　　　　　d. bevel gear

2. The differential is connected with the _____.

    a. universal joint　　　　　　b. propeller shaft

    c. final drive　　　　　　　　d. wheels

3. The differential assembly uses the _____ rotation to transfer power to the axle shaft.

    a. steering wheel　　　　　　b. steerable wheel

    c. drive shaft　　　　　　　　d. universal joint

4. The _____ is a gear assembly in a motor vehicle which allows the drive shaft to turn the wheels at different speed when the vehicle is turning around corner.

    a. clutch　　　　　　　　　　b. brake

    c. gearbox　　　　　　　　　d. differential

Exercise Three

Answer the questions:

1. What is the function of the power train?
2. Which components does the differential assembly include?

Lesson 13

Final Drives

　　The final drive(Fig. 1-13-1) transfers power from the engine and transmission to the wheels that drive the car. The final drive takes power from the spinning drive shaft and transfers it 90°to make the drive wheels turn. The final drive must also divide the power between the two drive wheels. When the vehicle comes straight ahead, the power divides equally so that both wheels receive an equal amount of power and move an equal distance. When a car goes around a corner, the outside wheel must move a greater distance than the inside wheel. Thus, the drive wheels must move at different speeds and with different driving forces.

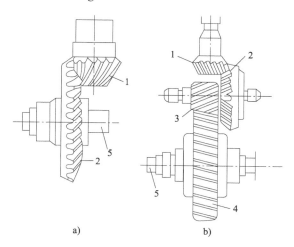

Fig. 1-13-1 The drive pinion and ring gear of the final drive

1-driving bevel gear(drive pinion); 2-driven bevel gear(ring gear); 3-driving cylindrical gear; 4-driven cylindrical gear; 5-semi-axle

If the two drive wheels moved at equal speeds, the wheels would skid as the car rounded the corner. This would cause the tires to wear rapidly. To avoid such wear, the final drive has a set of gears that can divide the power between the driver wheels. Thus, the wheels can turn at different speeds, if necessary. Because the gears allow a speed differential between the two drive wheels, the gear mechanism is called the differential. Some people confuse the differential with the final drive. The differential is really only the power-dividing gears, while the final drive includes the differential plus other parts.

Dividing power unequally is difficult. It becomes even harder when one wheel loses traction, such as on ice or slippery pavement. ① The final drive assembly tends to transmit power to the wheel that spins and no power to the wheel that may have normal traction. Most differential do not compensate for this situation and are called standard, open, or unlimited-slip differentials.

However, a more complex design avoids this problem. This type of final drive allows a certain amount of slip between the two drive wheels. When one drive wheel starts spinning at a much greater speed than the other wheel, the differential "senses" the change. ② The differential then transfers the power away from the wheel that spins and toward the one that does not spin. This type of differential is called a limited-slip differential.

The final drive gears provide the means for transmitting transmission output torque to the differential section(Fig. 1-13-2). The driver pinion, ring gears and differential assembly are normally located within the transaxle housing. There are four common configurations used as the final drives: helical, planetary, hypoid, and chain drive.

The drive pinion gear is connected to the transmission's output shaft and the ring gear is attached to the differential case. The drive pinion and ring gear assembly provide for a multiplication of torque.

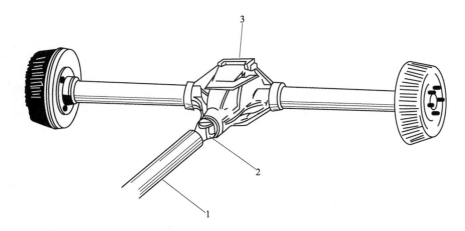

Fig. 1-13-2 The final drive transfers power from the drive shaft to the drive wheels
1-drive shaft;2-universal joint;3-final drive assembly

New Words

1. confuse [kən'fjuːz]   vt.   使混乱,混淆
2. pavement ['peivm(ə)nt]   n.   路面,铺筑材料
3. compensate ['kɔmpenseit]   vt.   赔偿,补偿
4. spin [spin]   vt.   使旋转
5. traction ['trækʃ(ə)n]   n.   牵引,牵引力
6. differential [ˌdifə'renʃ(ə)l] n.   差速,微分,差速器
                                adj.   差动的,微分的,差速的
7. transaxle ['træsæksl]   n.   驱动桥
8. hypoid ['haipɔid]   adj.   准双曲面齿轮的

Phrases and Expressions

1. final drive   主减速器,最终传动
2. limited-slip differential   防滑差速器
3. drive pinion   (主减速器)主动轮
4. ring gear   (主减速器)被动轮

Notes

①The final drive assembly tends to transmit power to the wheel that spins and no power to the wheel that may have normal traction.
主减速器总成趋向于将动力传到滑转的驱动轮,而不传到能正常产生驱动力的驱动轮。

②The differential then transfers the power away from the wheel that spins and toward the one that does not spin.
差速器不把动力传递给旋转的驱动轮,而传递给不旋转的驱动轮。

Part 1 Engineering Machinery Bases

Exercise One

Put the following expressions into Chinese:
1. final drive
2. limited-slip differential
3. drive pinion
4. ring gear
5. horizontal members of the chassis
6. longitudinal members of the chassis

Exercise Two

Choose one of the four words or phrases that best matches the definition according to the text.

1. When the vehicle comes straight ahead, _____ divides the power equally so that both wheels receive an equal amount of power.
   - a. clutch
   - b. differential
   - c. gearbox
   - d. universal joint

2. The drive pinion gear is connected to transmission' _____ and the ring gear is attached to the differential case.
   - a. input shaft
   - b. output shaft
   - c. lay shaft
   - d. idle shaft

3. The _____ takes power from the spinning drive shaft and transfers it 90° to make the drive wheels turn.
   - a. planetary gears
   - b. differential gears
   - c. drive pinion and ring gear
   - d. cylindrical gears

4. The final drive includes the _____ plus other parts.
   - a. clutch
   - b. transmission
   - c. brake
   - d. differential

Exercise Three

Answer the questions:
1. What components does the final drive include?
2. What is the limited-slip differential?

Lesson 14

Brake System

The most vital factor in the running and control of the modern vehicles is the braking system.

In order to bring the moving motor vehicle to rest or slow down in the shortest possible time, the energy of motion possessed by the vehicle must be converted into some other form of energy. The rate of slowing down or retardation is governed by the speed of conversion of energy. Kinetic energy is the energy of motion which is converted into heat given up to air flowing over the braking system.

The means of slowing down or bringing to rest a moving vehicle in the shortest possible distance is called brakes. Brake is a friction device for converting the power or kinetic energy of the moving vehicle into heat by means of friction.

The brake system comprises the brake assemblies and means to apply the brakes. The modern automobiles use two types of brake assemblies: wheel brakes and drive shaft brakes.

There are three types of wheel brakes: the drum brakes, the disc brakes and band brakes (commonly used in crawler machines).

The wheel brake, as the name implies, is located in the wheel. Fig. 1-14-1 is shown a drum brake assembly controlled hydraulically. It consists of the brake drum, an expander, pull back springs, a stationary backplate, two shoes with friction linings, and anchor pins. The stationary back plate is secured to the flange of the axle housing or to the steering knuckle. The brake drum is mounted on the wheel hub. There is a clearance between the inner surface of the drum and the shoe lining.

When the drum brake pedal is depressed, the rod pushes the piston of the brake master cylinder which presses the fluid. The fluid flows through the pipelines to the power brake unit and then to the wheel cylinders. The fluid pressure expands the cylinder pistons thus pressing the shoes to the drum. If the pedal is released, the piston returns to the initial position, the pull back springs retract the shoes, the fluid is forced back to the master cylinder and braking ceases. The hand brake must also be able to stop the car in the event of the foot brake failing. For this reason, it is separate from the foot brake and uses cable and/or rods instead of the hydraulic system.

Recently, most vehicles are mounted anti-lock brakes. The most efficient braking pressure takes place just before each wheel lock up. When you slam on the brakes in a panic stop and the wheels lock up, causing a screeching sound and leaving strips of rubber on the pavement, you do not stop the vehicle nearly as short as it is capable of stopping. Also, while the front wheels are locked up, you loose all steering control so that, if you have an opportunity to steer around the obstacle, you will not be able to do so. If the rear wheels are locked up, the vehicle will tend to spin out. Another problem occurs during an extended skid so that you will burn a patch of rubble off the tire.

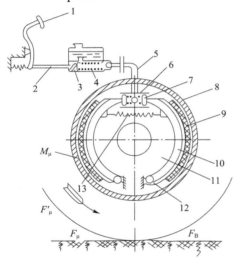

Fig. 1-14-1 Operation of braking system
1-braking pedal; 2-push rod; 3-master-cylinder piston; 4-master cylinder; 5-hydraulic lines; 6-wheel cylinder; 7-wheel cylinder pistons; 8-brake drum; 9-friction linings; 10-brake shoes; 11-back plate; 12-anchor pins; 13-pull back springs

Anti-lock brake systems solve this lockup problem by rapidly pumping the brakes whenever the system detects a wheel that is locked up. In most cases, only the wheel that is locked will be pumped, while full braking pressure stays available to the other wheels. This effect allows you to stop in the shortest amount of time while maintaining full steering control even if one or more wheels are on ice.

The system uses a computer to monitor the speed of each wheel. When it detects that one or more wheels have stopped or are turning much slower than the remaining wheels, the computer sends a signal to momentarily remove and reapply your pulse the pressure to the affected wheels to allow them to continue turning. This "pumping" of the brakes occurs at ten or more times a second, far faster then human can pump the brakes manually.

The system consists of an electronic control unit, a hydraulic actuator, and wheel speed sensors at each wheel. If the control unit detects a malfunction in the system, it will illuminate an ABS warning light on the dash to let you know that there is a problem. If there is a problem, the anti-lock system will not function but the brakes will otherwise function normally.

New Words

1. vital['vait(ə)l]	*adj.*	极其重要的,必不可少的;非常的
2. possess[pə'zes]	*vt.*	具有,拥有,占有
3. retardation[ˌriːtɑːˈdeiʃən]	*n.*	减速;制动,阻止
4. conversion[kənˈvəːʃ(ə)n]	*n.*	转变,转换;变换
5. contact['kɔntækt]	*n.*	接触,连接,联系
6. expander[ikˈspændə]	*n.*	膨胀器,扩张器
7. backplate['bækˌpleit]	*n.*	后板,后挡板
8. flange[flæn(d)ʒ]	*n.*	凸缘;法兰,法兰盘
9. retract[riˈtrækt]	*vi.*	缩回;缩进;收回
10. actuate['æktʃueit]	*vt.*	开动,操纵
11. leverage['liːv(ə)ridʒ]	*n.*	杆系;杠杆系;杠杆机构
12. event[iˈvent]	*n.*	事件,过程,会场
13. cease[siːs]	*vt.;vi.*	停止
14. fluid['fluːid]	*n.*	液,流体

Phrases and Expressions

1. drum brake	鼓式制动器
2. disc brake	盘式制动器
3. band brake	带式制动器
4. wheel brake	车轮制动器
5. drive shaft brake	驱动轴制动器
6. anchor pins	支撑销,副连杆销
7. master cylinder	主缸
8. panic stop	紧急停车
9. kinetic energy	动能

Exercise One

Translate the following expressions into Chinese:

1. parking brake system
2. drum brake
3. disc brake
4. band brake
5. braking system efficiency
6. anti-lock brake system

Exercise Two

Choose one of the four words or phrases that best matches the definition according to the text.

1. The means of slowing down or bringing to rest a moving vehicle in the shortest possible distance is called _____.
   a. steering   b. dumping
   c. braking   d. driving
2. There are _____ types of the wheel brakes.
   a. one   b. two
   c. three   d. four
3. There is a _____ between the inner surface of the drum and shoe lining.
   a. cooling water   b. lubrication oil
   c. sealant   d. clearance
4. _____ solve this lockup problem by rapidly pumping the brakes whenever the system detects a wheel that is locked up.
   a. anti-lock braking system   b. parking braking system
   c. hand brake   d. foot brake

Exercise Three

Answer the questions:

1. List the components that compose the anti-lock braking system?
2. What is the purpose of the braking system?

Lesson 15

Steering System

The function of the steering system is to convert the rotary movement of the steering wheel in the driver's hands into angular turn of the front wheels, and to multiply the driver's effort by leverage or mechanical advantage or booster so as to make it fairly easy to turn the wheels. The steering system includes a steering gear and a steering control linkage. The basic types of steering gears are recirculating ball type and rack-pinion type and worm-roller type.

They may be operated manually or with the aid of hydraulic power. The rotation of the steering wheel is imparted to the levers and rods of the linkage by which the steerable wheels are turned.

The steering gear increases the force transmitted from the steering wheel to the pitman arm and thus facilitating the turning of the steerable wheels. [1]Some automobiles are equipped with a hydraulic power steering system intended to decrease the efforts spent by the driver to turn the wheels and to damp the road jolts transmitted to the steering wheel. The booster is integrated with the steering gear(Fig. 1-15-1).

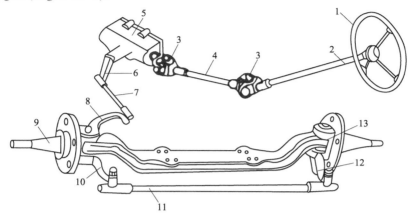

Fig. 1-15-1 Steering system

1-steering wheel;2-steering column;3-universal joint;4-steering shaft;5-steering gear;6-pitman arm;7-drag link;8-steering knuckle arm;9-left steering knuckle;10、12-steering arms;11-tie rod;13-right steering knuckle

The steering gear includes a housing, a worm, a ball nut, a rack made integral with the piston, a hydraulic booster, and a toothed sector integral with pitman arm shaft.

The ball nut of the steering gear is rigidly secured inside the rack-piston. To decrease friction between the worm and the nut, balls recirculate in their thread(Fig. 1-15-2). The rack-piston has resilient split rings of cast iron ensuring its tight fit in the steering gear housing. The rotation of the steering shaft is converted into progressive motion of the rack piston owing to displacement of the nut along the worm. The rack piston teeth turn the sector, hence the shaft with the pitman arm.

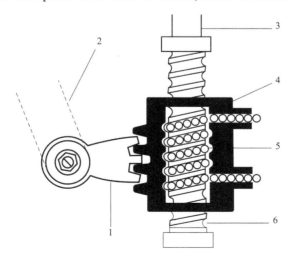

Fig. 1-15-2 Parts of recirculating ball steering gear

1-sector gear;2-pitman arm;3-steering shaft;4-ball nut rack;5-recirculating ball bearings;6-worm gear

The effort applied to the steering gear for cornering should be transmitted to the front wheels. This is performed by the steering linkage consisting of pitman arm, a tie rod, a drag link, a steering (upper) arm and a left and right knuckle arms. The steering linkage is so designed as to prevent side-slip of all automobile wheels in cornering, thus ensuring ease of steering and minimizing wear of tires. For this purpose it is necessary that the geometrical axes of all the wheels intersect in one point which is the common center of the circles described by the wheels.

The double-acting (two-chamber type) vane pump of the hydraulic booster with reservoir and filter is mounted on the engine. It is driven by a V-belt from the crankshaft pulley or by a gear train. The pump is connected to the control valve by two hoses: a pressure hose through which the hydraulic fluid is delivered from the pump and a return hose for returning the hydraulic fluid to the reservoir.

New Words

1. steering ['stiriŋ]	n.	转向;转向机构,操纵机构
2. angular ['æŋgjulə]	adj.	角度的;角的,尖的,倾斜的
3. booster ['buːstə]	n.	增压器;调压器,放大器
4. multiply ['mʌltiplai]	vt.;vi.	增大;放大,乘
5. leverage ['liːv(ə)ridʒ]	n.	杠杆机构,杠杆作用
6. advantage [əd'vɑːntidʒ]	n.	效益,利益,优点
7. impart [im'pɑːt]	vt.	给予;赋予,告诉
8. facilitate [fə'siliteit]	vt.	使便利;简化,促进,助长
9. damp [dæmp]	n.	潮气;雾气;湿气
	vi.	(使)潮湿,减弱;衰减,减振
10. jolt [dʒəult]	n.;vi.	振动;颠簸
11. thread [θred]	n.	螺纹,线(状物),丝(状体)
12. resilient [ri'ziliənt]	adj.	有弹性的;弹性的
13. split [split] (split, spliting)	vt.	劈开,割裂;分裂
	adj.	裂开的,分散的
	n.	分裂;割裂
14. vane [vein]	n.	叶片;导向叶片,(复)导向器
15. filter ['filtə]	n.	过滤,过滤器;滤清器
	vt.;vi.	过滤;滤清
16. pulley ['puli]	n.	滑轮;皮带轮;引轮
17. reservoir ['rezəwɑː(r)]	n.	容器,油箱,储气箱
	vt.	储藏,积蓄
18. globoid ['gləuboid]	adj.	球形的;球状的
	n.	球状体
19. knuckle ['nʌk(ə)l]	n.	关节,铰链,万向接头

20. twist [twist]	n.	曲解;扭曲	
	vt. ;vi.	拧;扭	
21. intermediate [ˌintəˈmiːdiət]	adj.	中间的	
22. tie [tai]	n.	结,扣;联系,联结杆	
	vt. ;vi.	系;结住	

Phrases and Expressions

1. steering gear —— 转向机构,转向器
2. mechanical advantage —— 机械效益
3. toothed sector —— 齿扇
4. rack-piston —— 齿条柱塞,齿条活塞
5. split ring —— 开口环,裂环
6. owing to —— 由于,因为
7. double-acting vane pump —— 双作用式叶片泵
8. gear train —— 齿轮系,齿轮链
9. triple roller and bearing —— 三重滚子轴承
10. adjusting screw —— 调整螺钉
11. pitman arm —— 转向垂臂
12. ball nut —— 球螺母
13. tie rod —— 转向横拉杆
14. drag link —— 转向纵拉杆
15. knuckle arm —— 转向节臂
16. recirculating ball steering gear —— 循环球式转向器

Notes

①Some automobiles are equipped with hydraulic power steering system intended to decrease the efforts spent by the driver to turn the wheels and to damp the road jolts transmitted to the steering wheel.

有些汽车装有液压转向系统,用来减轻驾驶员转向时的操纵力和减小传递到转向盘上的振动。

Exercise One

Put the following expressions into Chinese:
1. steering wheel
2. steering gear
3. steering system sensibility
4. steerable wheel
5. turning radius
6. steering knuckle

Exercise Two

Choose one of the four words or phrases that best matches the definition according to the text.

1. The rotation of the _____ is converted into prograssive motion of the rack-piston owing to displacement of the nut along the worm.

  a. steering system   b. steering shaft

  c. steerable wheel   d. driving wheel

2. The _____ is a device for converting the rotary motion of the driver's steering wheel into the angular turning of the front wheels.

  a. steering gear   b. steering wheel

  c. steerable wheel   d. reduction gear

3. The steering gear increases the force transmitted from the _____ to the pitman arm.

  a. steerable wheel   b. steering wheel

  c. reduction gear   d. modulator

4. The _____ is integrated with the steering gear.

  a. reduction gear   b. booster

  c. filter   d. modulator

Exercise Three

Answer the questions:

1. What is the function of the steering system?

2. What is the function of the steering gear?

Lesson 16

Frame and Suspension System

The frame is load-carrying beam structure consisting of tough steel sections welded, riveted, or bolted together. It forms a foundation for the car body and the parts of the several systems. The frame transmits the load through suspension system and axles to the wheels. It withstands the static and dynamic loads within permissible deflection. It should be stiff and strong to resist the severe twisting and bending forces to which it is subjected during motion of the vehicle on the road.

The frame is fabrication of the box, tubular, channel, angle section etc. The cross members reinforce the frame and also provide support for the engine and wheels.

The function of the suspension system (Fig. 1-16-1) is to absorb the road shocks and to prevent them from being transmitted to the other components of the vehicle. They protect the components from impact and dynamic load. Due to this provision, other components become safe during working and their life is increased. The occupants of the vehicle are not subjected to the jerks due to the presence of the suspension system and their journey becomes much smoother and less tiring. The suspension system maintains the stability of the vehicle during pitching or rolling while in motion.

Springs, torsion bar and other components necessary for jointing purpose are called the members of suspension system.

Part 1　Engineering Machinery Bases

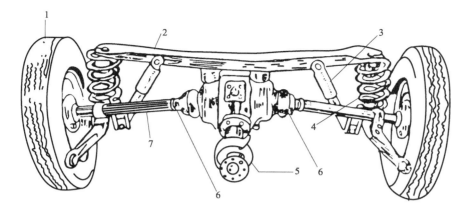

Fig. 1-16-1　Independent rear suspension

1-tyre;2-cross member;3-shock absorber;4-coil spring;5-center bearing;6-universal joints;7-half shaft

Leaf springs are made from flat strips of spring steel. Several strips are placed one on the other. They are jointed together by clamps and a central bolt. The length of each leaf decreases so that the spring assembly acts as a flexible beam and is of uniform strength. The longest strip is called main leaf spring or master leaf spring.

The coil spring is made from a special spring steel wire. The spring is generally circular in cross section and of suitable diameter to have the desired stiffness. The wire is wounded in the shape of a coil. The spring is formed at high temperature, cooled and proper heat treatment is given to it in order to have the characteristics of elasticity. When the wheel of the automobile experiences bump on the roadside, the coil spring compresses to absorb the shock energy.

The torsion bar performs the spring action by its resistance to twisting. Coil springs absorb force by compressing and leaf springs absorb force by bending. A torsion bar absorbs force by twisting. The bar is made from steel. The bar is mounted transversely in some of the vehicles, whereas in other constructions, it is employed lengthwise along the frame. One end of the torsion bar is fixed to the frame and it cannot rotate. The opposite end is mounted and subjected to twisting moment. The mounting enables the bar to twist at free end. A support arm on the twisting end is used to carry the wheel.

Springs alone are not satisfactory for suspension system. The ideal spring for automotive suspension would be one that would absorb road shock rapidly and then return to its normal position slowly. This is quite impossible in the case of springs, as they are by nature to oscillate. The spring must be a compromise between flexibility and stiffness. ①An extremely flexible spring, or too soft, would allow too much movement, while a stiff or hard spring would give too rough a ride. However, satisfactory riding qualities are obtained by using a fairly soft spring in combination with a shock absorber.

The most common in use on automobile at present is a hydraulic shock absorber which operates on the principle that fluids are incompressible and the resistance to the movement of the fluid through the restricting orifices imposes a drag on the spring movement, thus quickly damping out spring oscillations.

New Words

1. suspension [sə'spenʃ(ə)n]　　　n.　　　悬挂;悬挂物,悬置
2. torsion ['tɔːʃ(ə)n]　　　n.　　　扭转;扭力;转矩
3. helical ['helik(ə)l]　　　adj.　　　螺旋形的;螺旋(线)的
4. joint [dʒɔint]　　　n.　　　接合,接口
　　　　　　　　　　　vt.　　　连接,接合
　　　　　　　　　　　vi.　　　黏合
5. impact ['impækt]　　　n.　　　碰撞,冲击
　　　　　　　　　　　vt.　　　冲击,碰撞,装紧;压紧
6. dynamic [dai'næmik]　　　adj.　　　动力的,动态的,不断变化的
7. provision [prə'viʒ(ə)n]　　　n.　　　准备,预备,预防,供给
8. occupant ['ɔkjup(ə)nt]　　　n.　　　占有者,使用者,车内的人
9. jerk [dʒəːk]　　　n.　　　猛然一拉,振动,冲击
10. pitch [pitʃ]　　　vt.;vi.　　　纵向摇动,前后颠簸
　　　　　　　　　　　n.　　　投掷,纵摇,节距;螺距(机);沥青,树脂
11. elasticity [elæ'stisiti]　　　n.　　　弹性;弹力,伸缩性
12. transversely [trænsˈvəːsli]　　　adv.　　　横断地,横切地
13. twisting ['twistiŋ]　　　n.　　　扭曲;扭转
14. oscillate ['ɔsileit]　　　vt.;vi.　　　振荡;(使)来回振动
15. compromise ['kɔmprəmaiz]　　　n.　　　妥协,折中
16. orifice ['ɔrifis]　　　n.　　　孔;小孔,口
17. impose [im'pəuz]　　　vt.　　　强制
18. damp [dæmp]　　　n.　　　潮气,湿气,水蒸气,阻尼
　　　　　　　　　　　vi.　　　(使)潮湿,减弱,缓冲;减振

Phrases and Expressions

1. shock absorber　　　减振器
2. torsion bar　　　扭杆
3. helical coil spring　　　螺旋弹簧
4. be subjected to　　　受到……,经受,承受
5. heat treatment　　　热处理
6. in the case of　　　万一,遇到……时
7. the leaf spring　　　钢板(叶片)弹簧
8. the coil spring　　　螺旋弹簧

Part 1　Engineering Machinery Bases

Notes

①An extremely flexible spring, or too soft, would allow too much movement, while a stiff or hard spring would give too rough a ride.

一个太软或挠度极大的弹簧会产生过大的位移；而一个太硬或刚度极大的弹簧,行驶的平顺性又会太差。

while 连词,"而"。

Exercise One

Put the following expressions into Chinese.

1. leaf spring
2. coil spring
3. torsion bar
4. chassis
5. air spring
6. suspension system mechanics

Exercise Two

Choose one of the four words or phrases that best matches the definition according to the text.

1. The framework of a vehicle without a body and finders is referred to as _____.
   a. mounting   b. chassis
   c. driver's cabin   d. wagon
2. The _____ is the shaft or shafts of a vehicle upon which wheels are mounted.
   a. bolts   b. support
   c. axle   d. frame
3. The _____ is a strong beam, usually of iron or steel and supports the smaller beam.
   a. bracing   b. girder
   c. channel bar   d. panel
4. The _____ is load-carrying beam consisting of tough steel sections welded, riveted, or bolted togethe.
   a. tyre   b. wheel
   c. frame   d. shock absorber

Exercise Three

Answer the questions:

1. List the spring types of the suspension system?
2. What are the purposes of the frame?

Unit 2 Reference Translations and Answers

第一课 发动机结构与工作原理

内燃机是通过燃烧其汽缸内部的燃料而工作的发动机。燃烧汽油的内燃机是汽油机。

其他种类的内燃机依靠燃烧重油或燃料来进行工作。这些类型中应用最为广泛的当属柴油机。每一台发动机都是由主要的工作装置组成,通过辅助装置把工作装置组装在一起完成工作。

发动机包括两大机构,四(五)大系统。两大机构是:

1. 曲柄—连杆机构

(1) 发动机缸体;

(2) 汽缸;

(3) 汽缸盖;

(4) 活塞;

(5) 连杆;

(6) 曲轴;

(7) 飞轮;

(8) 曲轴箱。

2. 气门机构

(1) 气门;

(2) 挺杆;

(3) 推杆;

(4) 摇臂;

(5) 凸轮轴;

(6) 气门正时齿轮。

发动机包括的系统为:

(1) 燃油系统;

(2) 润滑系统;

(3) 冷却系统;

(4) 启动系统;

(5) 点火系统。

所有发动机都将能量从一种形式转化为另一种形式。工作中的内燃机在燃料燃烧期间,把储存在燃料里的化学能转化为热能。燃气的膨胀使其推动连接在曲轴上的活塞,这一过程使热能转化为机械能。

发动机将燃气膨胀所产生的能量转变为有用功,其结构如图 1-1-1 所示。

施加在活塞上的力通过连杆传到曲轴从而使其旋转。①活塞向下运动时,曲轴转半圈。②飞轮与曲轴相连,用来储存能量。飞轮的动量使活塞始终保持运动平衡,直到活塞接收到下一个能量脉冲。这样,活塞的往复运动即被转化为曲轴的回转运动。

完成整个工作循环的活塞的行程数随着发动机类型而不同。一个工作循环通常持续活塞的四个行程或曲轴转两圈。这种发动机被称为四冲程发动机或四冲程循环发动机。

在四冲程发动机中,这四个行程被称为进气行程、压缩行程、做功行程和排气行程。

进气行程:这一行程中,活塞通过曲轴向下运动。新鲜混合气通过打开的进气门被吸进汽缸。

压缩行程:燃气的压缩,点火,以及大部分的燃烧都在活塞接下来的向上的行程中进行。

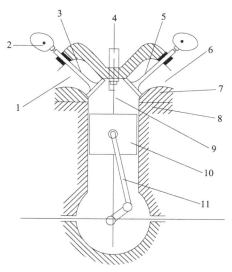

图 1-1-1 发动机结构
1-进气道;2-凸轮;3-进气门;4-火花塞;5-排气门;6-排气道;7-汽缸盖;8-缸体;9-燃烧室;10-活塞;11-连杆

③混合气在压缩时被点燃(柴油机则将燃料喷射在新鲜空气中),当活塞位于上止点时,燃烧进行过半。注意:在此行程中,进气门和排气门均处于关闭状态。

做功行程:燃烧所产生的热量使气体发生膨胀,从而使汽缸和活塞受到压力。在这个推力的作用下,活塞向下运动。这个行程的大部分阶段,进、排气门也均处于关闭状态。

排气行程:当排气门打开时,由于自身的膨胀大部分燃烧的废气排出汽缸。活塞通过向上运动将残余气体从排气门推出。

Key to the Exercises

Exercise One
1. 内燃机
2. 水冷发动机
3. 做功冲程
4. 曲轴
5. 缸壁磨损
6. 凸轮轴

Exercise Two
1. c   2. d   3. b   4. c

Exercise Three

1. suction, compression, power and exhaust stroke.

2. to convert the chemical energy stored in their fuels into heat energy during the burning part of their operation.

第二课　压燃式发动机

　　四冲程压燃发动机的压缩比是 16~24:1，并在压缩行程结束的时候把吸进工质的压力提高到 3~5①MPa，温度提高到 500~800℃。使用的燃油自燃温度大约是 400℃，所以当燃油以 10~25MPa 的压力喷射到燃烧室时就着火了。

　　在汽油发动机中，由化油器提供的、混合均匀的可燃混合气由火花塞点燃。但是在压燃式发动机（图 1-2-1）中，为了完全燃烧混合气在汽缸中彻底混合是在 35°~40°的曲轴转角范围内（喷油提前角）完成的。这个混合过程是非常困难的，并且要通过穿透力强的、雾化良好的喷雾和气体的涡流运动来完成，即使这样也只有 80% 的空气被利用，但火花点燃式发动机可以消耗吸进去的所有的空气。

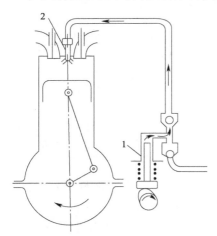

图 1-2-1　压燃式发动机结构
1-喷油泵；2-喷油器

　　与火花点燃式发动机相比，（柴油发动机）燃烧压力更高，需要强度更高、重量更大、精度更高的结构，②同时油泵、喷油器的精密加工也增加了它的成本。即使使用延伸性很好的铝合金，功率/重量也低于汽油机。由于更高的压缩压力以及燃油着火所需更高的热量，所以柴油机起动困难、怠速不稳、加速性差、③最高转速低。另外，不完全燃烧还会导致④动力性下降、排出有毒气体。

　　另一方面，压缩比大使柴油机热效率更高、经济性⑤更好。另外，柴油更便宜、着火的危险性更小，低速时转矩更大，实际工作中基本不需要维护。如果发动机正常工作基本不会积炭。

Key to the Exercises

Exercise One

1. 油环

2. 气环

3. 压燃式发动机，柴油机

4. 火花点燃式发动机，汽油机

5. 曲柄销（连杆轴颈）

6. 气门机构

Exercise Two

1. d 2. a 3. b 4. a

Exercise Three

1. oil control ring and compression ring
2. to convert rotary motion into reciprocating motion

第三课　柴油发动机供油系统

柴油发动机可用燃油类型从高挥发性的喷射燃油、煤油到炉用重油。柴油发动机使用不同类型燃油工作情况取决于发动机的工作条件和燃油特性。

柴油是汽油馏分从石油中蒸馏之后煤油、汽油和太阳油馏分的混合物。柴油最重要的性能是用十六烷值评价的着火性、黏度、凝点和纯净度等。柴油根据凝度、闪点和黏度值分为不同的等级。

柴油发动机燃油系统包括油箱、粗滤器、细滤器、带手注油器的输油泵、带调速器的喷油泵、喷油正时齿轮、带喷油器的喷油器体和低、高压油路(图1-3-1)。

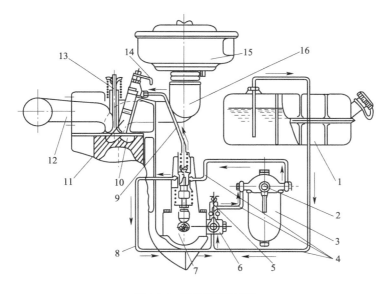

图1-3-1　柴油机燃油系统

1-油箱;2-溢流阀;3-滤清器;4-低压油路;5-手动注油泵;6-输油泵;7-喷油泵;8-回油管;9-高压油路;10-燃烧室;11-喷油器;12-排气管;13-排气门;14-排油管;15-空气滤清器;16-进气管

在发动机工作期间,输油泵把油从油箱中吸出经过粗滤器、再经过细滤器进入到喷油泵,通过高压油路燃油被泵到喷油器,根据发动机的着火顺序雾化的燃油喷到汽缸中,多余的燃油从喷油泵和喷油器返回到油箱。空气经过空滤器进入汽缸。

根据特定的顺序喷油泵在高压作用下把燃油喷到汽缸中,喷油泵安装在汽缸排之间通过齿轮传动由凸轮轴驱动。

泵包括泵体、凸轮轴、分泵(柱塞套筒偶件与汽缸数相同)、柱塞控制机构。[①]泵的前端安装了一个刚性调速器,它根据负荷来计量燃油喷射量从而保持由司机设定的发动机的转速。喷油泵凸轮轴的后端安装喷油正时齿轮,它根据发动机转速来改变发动机喷油时间。

喷油泵的分泵(柱塞套筒偶件)包括柱塞、套筒、滚子挺杆和出油阀。套筒在不同的高度设有两个油道,柱塞上部有两个油道和一个斜槽,柱塞安装在套筒内(成一偶件)。

②当柱塞在弹簧作用下向下移动时,输油泵产生的压力使燃油通过泵体的纵向进油道充满柱塞顶部的所有空间。当柱塞在凸轮和挺杆作用下向上移动时,燃油从出油道泄出直到柱塞(斜槽边)把套筒油道封住。当柱塞继续上行压力达到出油阀的极限压力时,柱塞略微抬起,燃油从高压油路进入到喷油器,柱塞保持向上移动建立的压力用来克服喷油器轴针弹簧压力,轴针抬起喷油开始。当柱塞(斜槽边)打开套筒油道时喷油停止,此时油压下降,出油阀的减压环带在弹簧作用下落座使喷油器与出油阀之间的油路容积增加,确保强制停止喷油。当齿条(横向)移动时,柱塞转动、斜槽提前或推后打开套筒油道,因此,油道打开时间(喷油时间)和喷入汽缸的油量都会改变。

喷油器的作用是把计量的、雾化良好的燃油在压力作用下喷进汽缸。

不同的发动机应选用最佳的喷油器,这就意味着最佳燃烧、最小污染排放和最大的功率输出。

闭式喷油器包括轴针式和孔式喷油器。轴针式喷油器常用在非直喷式发动机直列喷油泵上,孔式喷油器用在直喷式发动机直列喷油泵上。

闭式喷油器包括钢制喷油器体、螺母、喷嘴、轴针、轴和滤清器。燃油经过滤清器、垂直油道、环形槽进入到喷嘴处的燃油环隙。当环隙压力克服弹簧压力时,轴针从针座上抬起,燃油喷射到燃烧室。当油路压力下降时,轴针复位(喷油结束)。多余的燃油从旁道返回到油箱。

Key to the Exercises

Exercise One

1. 低高压油路
2. 柱塞套筒偶件
3. 喷油泵
4. 输油泵
5. 孔式喷油器
6. 轴针式喷油器

Exercise Two

1. a　　2. c　　3. b　　4. a

Exercise Three

1. ignitability , viscosity, pour point ,purity, etc.

2. fuel tank , a primary filter, a secondary filter, a fuel supply pump with a hand primer, an injector pump with a speed governor and automatic injection timing clutch, nozzle holders with nozzles, low and high pressure fuel lines

第四课　汽油发动机燃油喷射系统

单点喷射和多点喷射用于复杂的计算机控制的燃油喷射系统中,每种都采用周期性或定时喷射来控制燃油量。

单点喷射是指节气门体喷射,意味着燃油从一个地方吸进燃烧室。系统进气歧管类似于化油器发动机,但化油器被节气门体取代。

节气门体包括一个或两个电磁操纵的喷油器,把油直接喷在节气门片的上方(图1-4-1)。燃油在压力作用下提供给喷油器。节气门片像化油器一样由节气门联动装置控制。计算机给电磁操纵的喷油器提供电压脉冲信号,使喷油器开启(工作),把燃油喷进节气门腔内,喷油(持续)时间取决于由传感器感应的发动机工况。喷油器开启时间越长喷油越大。当发动机负荷和转速增加时,喷油器开启时间也会增加与增加的空气流量相匹配。喷油器开启时间等于脉冲宽度。脉冲宽度越长,喷油量越大。吸进的空气量由节气门的开度控制。传感器通常包括转速传感器、空气流量传感器、进气歧管压力传感器、节气门位置传感器、水温传感器、空气温度传感器和氧传感器。

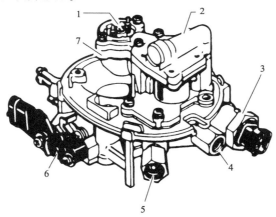

图1-4-1　节气门体喷射系统

1-喷油泵;2-燃油计量器壳;3-怠速空气控制阀;4-回油孔(通油箱);5-进油孔(通油泵);6-节气门位置传感器;7-燃油计量装置

电子燃油喷射系统是以极好的节气门开度反应和良好的驱动性能为特征。[①]但是,实践证明该系统最适用于进气歧管横截面小的发动机,这样发动机低速时仍能保持可燃混合气在歧管内高速流动,减小了混合气中较重的燃油颗粒从气流中离析的可能。

多点喷射是指气道喷射,也就是燃油从多点吸入到汽缸。每个进气道安装一个喷油器,燃油直接喷入到气道,也就是在进气门靠歧管一侧。多点喷射是目前研发的最先进的燃油控制形式,优点如下:

(1)[②]精确地将等量的燃油直接喷入每个汽缸的进气口,消除了原来已混合的空气和燃油通过进气管时燃油本质上分配不均的现象;

(2)燃油经过进气歧管时不会出现冷凝现象,所以不需加热空气或进气歧管;

(3)混合器低速经过进气歧管时燃油颗粒不会从气流中离析,所以歧管的横截面积可以

更大,高速时可以使容积效率(汽缸充气能力)更大;

(4)虽然在气道处会出现浸油现象,但绝大程度上可以避免歧管浸油现象。如果燃油从进气歧管吸入,尤其在低温的运行情况和加速情况下部分燃油会遗留在歧管底部或歧管壁上,燃油计量装置会计算出这部分燃油量防止混合气过稀。另外,还要计算由于混合气气化进入汽缸导致高真空情况下的燃油。

通常气道喷射的发动机会提供更好的动力性和驱动性,同时会保持或降低排放污染增加燃油经济性。

Key to the Exercises

Exercise One
1. 压缩的混合气
2. 曲轴位置传感器
3. 电子燃油泵
4. 以火花的形式放电
5. 容积效率
6. 高压电流

Exercise Two
1. b   2. c   3. a   4. b

Exercise Three
1. single point and multi-point injection
2. excellent throttle response and good driveability

第五课 发动机润滑系统

没有摩擦力汽车就不能移动。发动机内多余的摩擦意味着快速地损坏。内部摩擦不能消除但可以通过使用减少摩擦的润滑剂使它大大地减小。

汽车发动机中的润滑油有如下功能:

(1)通过润滑,减小发动机运动件之间的摩擦:

a)减小过多摩擦产生的具有破坏作用的热量;

b)储存克服过多摩擦浪费的能量。

(2)作为密封剂防止构件之间(混合气)的泄露,如活塞、活塞环和汽缸之间;

(3)通过在产生摩擦的构件之间流动带走热量;

(4)冲走由于摩擦表面磨损磨掉的金属碎屑。

现代汽车的润滑系可以分为"压力润滑"和"飞溅润滑",[①]不过,这两种润滑是以多种结合的方式来使用的。大多数的小客车都是使用通过机油泵的压力作用把机油压到绝大多数的旋转运动件和往复运动上的压力润滑系统。

飞溅润滑系统利用连杆一端的油匙穿过油底壳时把油溅到各种构件上,飞溅的机油通常变成油膜润滑构件。

Part 1　Engineering Machinery Bases

　　在压力润滑系统中机油由凸轮轴驱动的机油泵来提供,泵从油底壳中吸油再送到沿着②缸体纵向布置的主油道中。大多数的油道是在缸体上钻出来的孔,不过,许多发动机也使用钢管或铜管。主油道通过相连的油道(横向油道)直接把油提供给(曲轴)主轴承。在主轴承处,机油通过曲轴上钻出来的油孔被压到连杆轴承。在连杆轴承处,机油通过连杆上钻出来的油孔被压到活塞销。从(曲轴)主轴承处,分油道(横向油道)在压力作用下把油从主轴承送到凸轮轴。分油道中的一个油道可以通过喷射的机油润滑正时齿轮。

　　在这种设计中,连杆轴承上部的喷油孔用来润滑汽缸壁的承载面。

　　图1-5-1所示顶置气门发动机气门操纵机构润滑系统,该系统中与主油道垂直的支油道把机油提供给空心的摇臂轴。凸轮轴在每个气门处都有计量装置提供所需的机油。③汽缸盖上的油道使机油在重力的作用下返回到油底壳。即使在全液压润滑系统中,连杆轴承两端的机油也会被甩出用来润滑缸壁和凸轮,活塞油环控制机油分配使机油向上润滑缸壁,同时刮掉多余的机油。轴承两端溢出的机油和从活塞、汽缸流下的机油都返回到油底壳用来再循环。

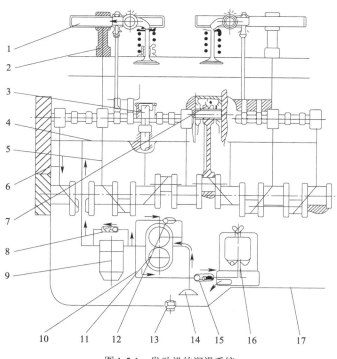

图1-5-1　发动机的润滑系统

1-摇臂轴;2-上油道;3-机油泵传动轴;4-主油道;5-横向油道;6-喷油嘴;7-连杆小头油道;8-机油粗滤器旁通阀;9-机油粗滤器;10-油管;11-机油泵;12-限压阀;13-放油塞;14-集滤器;15-机油细滤器限压阀;16-机油细滤器;17-油底壳

　　发动机机油(润滑油)分为两个基本类型:石油基机油和合成机油。石油基机油包括各种添加剂,所以实际上它们也是合成机油。抗磨添加剂可以为包括凸轮、活塞、汽缸壁等运动件抛光,这在新发动机磨合期和大修后都至关重要。防腐剂通过阻碍酸性成分的形成减小有害酸的形成,防腐剂在燃烧产物损坏发动机之前和它们进行中和反应。清洁—分散剂在发动机运行过程中清洁发动机构件并且使它们悬浮在机油中。黏度指数改善剂的作用是

67

在各种温度下稳定或改善机油的黏度,改善机油的油体和流动性。泡沫抑制剂减小泡沫形成的趋势,机油与空气混合所产生的热量和所需要的搅拌都会产生泡沫,泡沫会减小机油的润滑性能使发动机缺油出现故障。

Key to the Exercises

Exercise One

1. 防腐剂
2. 磨料
3. 润滑系统
4. 机油泵驱动轴
5. 最佳的润滑条件
6. 顶置气门发动机

Exercise Two

1. b   2. a   3. c   4. a

Exercise Three

1. pressure and splash lubrication
2. (1) Reducing friction between moving parts of the engine;
   a) Reducing amount of destructive heat generated by excessive friction;
   b) Conserving power that would otherwise be wasted in overcoming excessive friction.
   (2) By acting as a seal to prevent leakage between parts such as pistons, rings and cylinders.
   (3) By flowing between friction-generating parts to carry away heat.
   (4) Washing away abrasive metal worn from friction surfaces.

第六课　发动机冷却系统

　　汽车发动机汽缸内燃油/空气混合气的燃烧会产生热量。这些热量必须要转移以防止构件产生过多的膨胀、黏着和裂纹,所以,冷却系的作用是在任何转速、任何驱动状况下使发动机在最有效的温度下工作。

　　当燃油在发动机内燃烧时,三分之一的热量转换成能量,三分之一通过排气管排放掉,剩下的三分之一必须要靠冷却系处理。

　　最常见的方法是使用水(冷却液),把发动机热量先传递给水,然后冷空气吹水把热量传递给空气。

　　在图1-6-1所示的水冷发动机中水套环绕汽缸,水套是在缸体内铸出来的水道,冷却水由水泵泵出进入水套。

　　发动机水泵有各种类型,但大多数是离心式,①由旋转叶片组成,基本不用齿轮或柱塞式的容积排量式。水套中的热水通过上水管进入到散热器中,在泵回到下水管再次进入水套之前在散热器中由外边的空气冷却。

Part 1 Engineering Machinery Bases

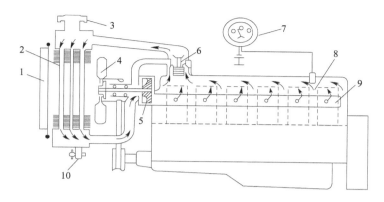

图 1-6-1 发动机的水冷系统
1-百叶窗;2-散热器;3-散热器盖;4-风扇;5-水泵;6-节温器;7-水温表;8-水套;9-分水管;10-放水阀

当发动机刚起动时不需要冷却,所以恒温器切断到散热器的水流。当发动机温度达到82~91℃时恒温器开始打开,使冷却水通过恒温器。当冷却液温度达到93~103℃时恒温器全开。

散热器是一种热转换器,把水中的热量传递到汽车外面的空气中,这样冷却水可以返回到水套中吸收更多的热量。散热器由上水箱、下水箱组成,它们之间是一系列的细管使水从上水箱流到下水箱,空气流经细管,使水中的热量散发到空气中。

散热器安装在汽车前部,目的是汽车的运动可以提供气流。当汽车静止时,风扇可以确保冷却过程的持续。风扇通常和水泵相接由发动机驱动,不过有时也由电动机驱动。

冷却水从水套经软管流进散热器,软管易损坏和裂纹,因此要定期检查,如果发现损坏及时更换。

Key to the Exercises

Exercise One
1. 冷却水循环
2. 发动机冷却系统
3. 水冷发动机
4. 空调系统
5. 风冷发动机
6. 曲轴箱换气

Exercise Two
1. b   2. b   3. c   4. d

Exercise Three
1. To control temperature of engine cooling water.
2. Radiator, fan, water pump, thermostat, water temperature gauge, water jacket, connecting pipes, drain tap, header tank, bottom tank.

第七课 发动机起动系统

起动操作简便是内燃机的主要性能特点之一。为起动发动机,也为了空气和燃料能够充分混合,并且可燃混合气能够得到充分的压缩和燃烧,让曲轴以足够的速度转起来是必要的。确保发动机能够顺利起动的曲轴的最低转速称为起动速度。起动速度取决于发动机类型和起动条件。

起动系统由机械部件和电子部件组成,两者配合工作来起动发动机。起动就是要将储存在电池内的电能转化为机械能。为完成这一转化,需要使用起动机(电动机)。化油器发动机(汽油机)的起动速度是 40~50r/min,柴油机是 150~250r/min。曲轴转速过低会使发动机起动困难,因为这样,汽缸内的气体会有更多的时间从接缝处泄漏,还会将其压缩所产生的热量散发到发动机的零部件上,这会导致在压缩行程结束时,汽缸内气体的压力减小,温度降低。

起动发动机时,让曲轴转起来是一件很费力的事,因为这一过程不仅要克服发动机中运动部件间的摩擦,还要克服汽缸内的气体被压缩时所产生的阻力。所使的力取决于发动机的温度,温度越低所需的力越大,因为润滑油的黏度越大。

柴油机和汽油机均不可自行起动。为了顺利起动,需借助外力使发动机曲轴旋转起来。内燃机的起动可用以下几种方法:

1)人力起动;
2)电力起动。

电动机起动是起动汽车发动机最常用的方法,随着起动电机的发展,完全电起动系统应运而生。例如,蓄电池用来为起动电机的供电,另外,发电机要给蓄电池充电补充起动机消耗的电能。

起动系统为起动内燃机提供能量,直到内燃机能够凭其自身能量进行运转。为了达到这个目的,起动电机从蓄电池得到电能,并把电能转化成机械能,机械能通过传动机构传到飞轮。

典型起动系统由五部分组成:蓄电池、起动开关、传动装置、起动机电磁线圈(控制装置)和起动电机(图1-7-1)。

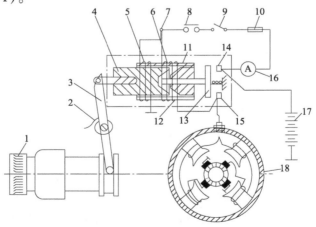

图1-7-1 电磁操纵式起启动系统

1-驱动齿轮;2-复位弹簧;3-拨叉;4-活动铁芯;5-保持线圈;6-吸拉圈;7-接线柱;8-起动按钮;9-起动总开关;10-熔断丝;11-黄铜套;12-挡铁;13-接盘;14、15-接线柱;16-电流表;17-蓄电池;18-直流电动机蓄电池的作用相当于一个能量源。它从充电系统接受能量并将其储存起来直到有需要。然后它以电流的形式向起动电路供电。起动电机是一种结构紧凑、动力很大的直流电机，可以快速起动发动机。它使(起动齿轮)小齿轮转动。传动装置的小齿轮与起动电机的端部相连。当驾驶员接通开关，小齿轮与飞轮环齿啮合，从而使曲轴转动并起动发动机。

起动系统操作简便，大大简化了驾驶员的工作强度，但需要有精湛的维护技能，还要有一定的能量储备，因此起动次数将会受到限制。[1]许多装有电驱动系统的发动机也提供人力起动装置，以防蓄电池或起动电机出现故障。

起动机电磁线圈(控制装置)是一个装在起动电机上面部的电磁开关。它有两个重要功能：控制蓄电池和起动电机间的电路；移动小齿轮与飞轮环齿啮合或脱离。

为了方便在低温条件下起动柴油机，还会用到减压装置和加热装置。

Key to the Exercises

Exercise One
1. 柴油发动机机飞轮环齿
2. 起动发动机的驱动齿轮
3. 电动马达
4. 压缩比
5. 起动机的电磁线圈
6. 旋转飞块的离心力

Exercise Two
1. c   2. d   3. b   4. a

Exercise Three
1. Combining mechanical and electrical parts that work together to start the engine.
2. Hand starting and electric starter motor.

第八课 汽油发动机点火系统

点火系统是电器系统的一部分，当燃烧室产生用来点燃混合气火花时，点火系统将电流传送到火花塞。

为了产生高压电来击穿系统设定的火花塞间隙，人们采用过很多方法。所有这些方法都以电磁感应原理为基础的。

通常，点火系统有三种：传统的蓄电池(触点式)点火系统、电子点火系统和微机控制点火系统。传统的蓄电池点火系统在汽车上的应用已超过60年。从20世纪60年代起，制造商们就已经开始将电子点火系统应用到高性能汽车上了。然而直到20世纪70年代早期，固态(晶体管)系统才出现在国内客车上。由于更为严格的废气排放控制标准和改善燃料经济的需求，这些系统目前在应用上处于统治地位。

汽车点火系统有两个基本功能：第一，它必须控制火花（能量）以及火花塞点火的时间与发动机工况需求相匹配；第二，点火系统必须将蓄电池电压升高到高压值，这样才能克服火花塞的电阻来使火花塞点火。

现代点火系统依靠蓄电池来工作。传统点火系统由以下几部分组成：蓄电池、点火线圈、分电器、电容器、点火开关、火花塞、附加电阻器和必要的低高压线路。

图1-8-1所示的是六缸发动机上的蓄电池点火系统。

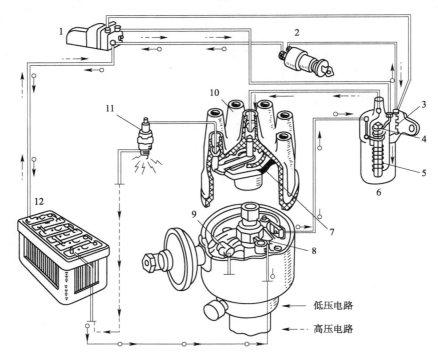

图1-8-1　蓄电池点火系统

1-启动开关；2-点火开关；3-附加电阻；4-初级绕组；5-次级绕组；6-点火线圈；7-分电器；8-断电器；9-电容器；10-分电器盖；11-火花塞；12-蓄电池

汽车点火系统分为两个电路：初级电路和次级电路。初级电路电压低。电路依靠蓄电池电流工作，电流由触点和点火开关控制。

次级电路由以下几部分组成：点火线圈的次级绕组、分电器和线圈间的高压导线、分电器盖、分电器转子（分火头）、火花塞导线和火花塞。

初级电路从蓄电池开始，经过电流表、点火开关、较粗的初级绕组（绕在软铁叠加的铁芯上），最后通过触点接地。电容器的一端与初级电路相连，另一端接搭铁。

次级线圈绕组在电路上不直接与初级电路相通。它从线圈的接地端开始，经过大约21000匝的细线绕组，然后通过绝缘效果较强的导线连接到分电器盖的中心电极，炭棒将电流送到分火头，分火头转动将电流分给六个侧电极，接着通过火花塞导线将电流输送到火花塞。当电流击穿火花塞间隙时点燃汽油混合气，最后接搭铁。

装有分火头的分电器轴在发动机凸轮轴的带动下旋转。在该轴的顶端，即分火头的下方，[①]设有一个六棱凸轮，每一棱配合图1-8-1所示发动机的六个汽缸之一进行工作（凸棱的数目与汽缸数相等,最多 8 个,对于一套触点不够用的十二缸和十六缸发动机而言,需要使用设有六棱和八棱凸轮的两套触点进行工作)。

　　在 20 世纪 70 年代早期,美国汽车开始安装电子点火系统,不仅满足严格的气体排放需求还能增加燃油经济性。电子点火系统不用触点作为开关装置来接通或断开初级绕组的电流,取而代之系统使用晶体管作为开关控制初级绕组的电流,因此,对于严格控制排放标准的发动机,它不仅可以提供点燃较稀混合气的所有的高压,还能击穿由于磨损间隙增加的火花塞。电子点火系统优点可以概括如下:

　　1) 尤其对于高速发动机可以产生更高的、有效的次级电压;

　　2) 在发动机任何转速下,保持可靠的、恒定的性能;

　　3) 对于变化的点火提前角反应会更敏捷;

　　4) 减小系统的维修成本。

　　微机控制点火系统结合了计算机技术,从单独的控制系统发展成为现代集中控制系统。该系统包括各种传感器、电子控制单元和点火执行器,精确地控制点火时刻、通电持续时间和爆震燃烧。

Key to the Exercises

Exercise One

1. 可燃混合气
2. 电子点火系统
3. 互感
4. 火花塞电极
5. 初级绕组和次级绕组
6. 火花塞

Exercise Two

1. b　　2. c　　3. a　　4. b

Exercise Three

1. Spark ignition.

2. Carrying the electrical current to the spark plug when the spark necessary to ignite the fuel air mixture in the combustion chamber is produced.

第九课　离　合　器

　　发动机产生动力驱动车辆。传动系把发动机的动力传递给车轮。传动系包括飞轮后部到车轮的所有构件。这些构件包括离合器、变速器、主传动轴和主减速器总成。

　　离合器是把主动件与被动件连接或断开的摩擦装置。应用在汽车上,离合器通常与飞轮连接使发动机与手动变速器平稳结合或断开。

车辆上有许多类型的离合器,分类如下:

1. 摩擦式离合器。

1)单片式;

2)多片式。

(1)湿式;

(2)干式。

2. 螺旋弹簧式离合器。

3. 膜片弹簧式离合器。

4. 刚性连接离合器。

1)凸牙式;

2)花键式。

因为内燃机在低速时不产生(或产生很小的)功率或转矩,所以在车辆行驶之前发动机必须达到一定的转速。但是,快速旋转的发动机突然与静止车辆的传动系连接就会产生强烈的振动。

①随着负荷的增加,发动机转速也会下降。因此,需要合理、平稳的起动。安装手动变速器的车辆合理、平稳的起动是靠机械离合器完成的。

离合器利用摩擦力工作。离合器的主要部件是压盘(主动盘)和被动盘。压盘和飞轮连在一起,被动盘安装在变速器的输入轴上。这样,压盘通过弹簧压向被动盘,所以通过摩擦力把转矩从发动机传递给变速器的输入轴。

离合器有两个由发动机刚性地驱动的构件和第三个与变速器轴连接的构件。当踩下离合器踏板使这些构件分离时,发动机转动而不带动变速器轴,因此齿轮可以轻易换挡,②发动机怠速,汽车停车而不熄火。离合器的摩擦表面设计成当压盘刚开始施加作用力时被动盘与主动盘打滑,当作用力增加时,被动盘的速度逐渐达到主动盘的速度。当三者速度相等时,打滑完全停止,③三个构件紧紧相连。

传动靠三个构件间的摩擦力来完成,摩擦力取决于接触材料和把它们压在一起的压力,压力靠弹簧保持并且必须足够大防止离合器完全接合时打滑,摩擦材料必须能够提供足够大的摩擦力来传递载荷。当踩下离合器踏板时,弹簧被压缩,发动机与变速器脱离接合。

图1-9-1所示为典型的离合器解剖图。主要构件是主动件、被动件和操作件。

主动件包括与铸铁压盘或主动盘连接的离合器壳,压力弹簧和分离杠杆。整个总成用螺栓连接在飞轮上并且始终和飞轮一起转动。飞轮作为第二个主动件和压盘一起在弹簧的作用下紧紧地把被动件夹在中间。为了散发工作过程中摩擦产生的热量,离合器壳开有通风口。

被动件包括被动盘,它可以沿着离合器轴的花键部分自由的轴向滑动,④同时也是靠这些花键齿驱动离合器轴。离合器被动盘的两个承载面均带有摩擦材料。

操纵机构包括制动踏板、连接件、分离轴承、分离杠杆和确保离合器正常工作的弹簧。

Part 1　Engineering Machinery Bases

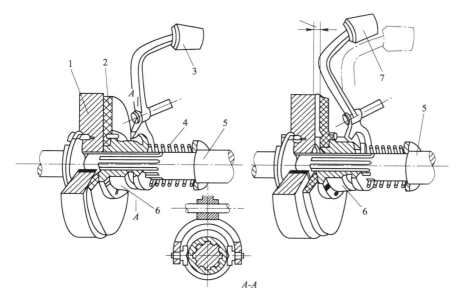

图 1-9-1　摩擦式离合器的工作原理
1-飞轮;2-被动盘;3、7-踏板;4-弹簧;5-被动轴;6-被动毂

Key to the Exercises

Exercise One
1. 发动机最大转矩
2. 发动机产生的功率
3. 离合器技术维护
4. 离合器的压力弹簧
5. 液压离合器
6. 离合器的自由行程

Exercise Two
1. c　　2. d　　3. b　　4. c

Exercise Three
1. In conjunction with an engine flywheel to provide smooth engagement and disengagement of the engine and manual transmission.
2. The clutch, the transmission, the drive shaft, the final drive assembly, axles and wheels.

第十课　变　速　器

变速器是安装在发动机与车辆驱动轮之间的用来改变车速和(动力)的装置。它可以改变发动机转速(转/分)与车轮行驶速度(转/分)之间的传动比来最好的满足特殊的驱动条件。变速器可以改变从发动机曲轴传递到主传动轴的转矩、改变车辆的行驶方向、长时间驻车或滑行时切断发动机与传动线路的动力。车辆起动或满载爬坡比它在水平路面上运行(惯性力大、行驶阻力小)所需要更大的转矩作用在车轮上。为了满足转矩变化的需求,需要使用特殊的齿轮箱,这种齿轮箱叫变速器。

根据传动比的不同,变速器可以分为三类:

1. 有级变速器:齿轮传动,具有若干个定值传动比。
1)固定轴齿轮式;
2)行星齿轮式。
2. 无级变速器:具有连续、可变的传动比。
1)液力式(传动部件是液力变矩器);
2)电力式(传动部件是直流串励式电动机)。
3. 综合式变速器:由液力变矩器和固定轴齿轮式变速器组成的液力—机械式变速器。

图 1-10-1 为滑动啮合三挡变速箱。变速箱由壳、输入轴和输入轴齿轮、输出轴和输出轴齿轮、惰轮轴、倒挡齿轮、(另外的)一组齿轮和换挡机构组成。

发动机转矩从离合器的被动盘通过离合器轴(第一运动轴)输出,离合器轴在变速器壳上安装的轴承内转动,和轴一体有一个小齿轮,小齿轮与下面副轴上相应的小齿轮永久的啮合。① 这对齿轮叫常啮齿轮。

副轴由四个齿轮组成,包括或者在固定轴上转动或者在变速器壳的轴承内转动的常啮齿轮。

花键轴(一端)支撑在离合器轴的导向轴承里,(另一端)支撑在后面的变速器壳的轴承中,通过万向节与主传动轴连接在一起。在花键轴上滑动的是两个带套筒的花键小齿轮,拨叉安装在套筒里。

最高挡或直接挡

二挡齿轮在面对离合器那面有凸牙,② 用拨叉在花键轴上滑动时凸牙与离合器轴上相应的凸牙匹配,在两个轴之间得到刚性、直接传动即直接挡(图 1-10-1d)。

二　　挡

对于二挡挡位,主轴齿轮与副轴二挡齿轮滑动啮合,传动是间接的,传动路线是从离合器轴经过常啮齿轮到副轴再通过二挡齿轮返回到主轴(图 1-10-1c)。

传动比是两对啮合齿轮的乘积,因此,

$$二挡传动比 = \frac{离合器转速}{主轴转速}$$

$$= \frac{离合器轴转速}{副轴转速} \times \frac{副轴转速}{主轴转速}$$

$$= \frac{副轴常啮齿轮齿数}{离合器常啮齿轮齿数} \times \frac{主轴二挡齿轮齿数}{副轴二挡齿轮齿数}$$

按照从主轴、副轴到离合器轴的顺序,根据相啮齿轮的齿数,下列公式更容易记住。

$$传动比 = \frac{③主轴齿轮齿数}{相应副轴齿轮齿数} \times \frac{副轴常啮齿轮齿数}{离合器轴常啮齿轮齿数}$$

Part 1 Engineering Machinery Bases

一　　挡

在这种情况下,主轴上的一挡齿轮与副轴上相应的一挡齿轮啮合,因为齿轮尺寸不同可以获得比二挡更大的减速比,传动比用一挡齿轮的齿数利用类似的方法计算(图1-10-1b)。

倒　　挡

副轴上的第四个齿轮永久的和倒挡惰轮啮合,倒挡惰轮在固定在变速器壳上的短轴上转动。当主轴一挡齿轮与惰轮滑动啮合时主轴的旋转方向反向。计算倒挡传动比时,只采用主轴一挡齿轮的齿数和副轴倒挡齿轮的齿数,因为惰轮齿数与传动比无关(图1-10-1e)。

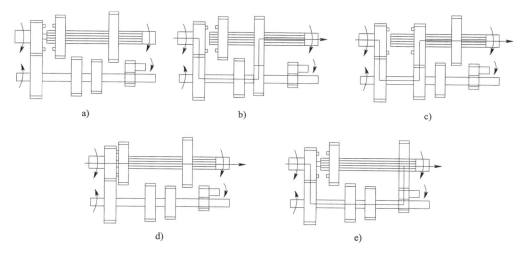

图1-10-1　滑动啮合3挡变速器
a)空挡;b)一挡;c)二挡;d)三挡;e)倒挡

当今自动变速器广泛用在现代汽车和工程机械上。它使驱动更加容易,包括三个基本系统——液力变矩器、变速齿轮系统和液压系统。

三个系统组装成一体,直接固定在发动机的背部。液力变矩器与手动变速器的离合器作用一样。液压系统是变速器的"智能系统",它必须判断什么时候换挡,通过提供必要的油压(加压或减压)使变速齿轮系统在正确的时间换挡。

液压系统用来感受(发动机)的转速和负荷,是一个像迷宫一样的、复杂的油道和油门系统。

变速齿轮系统由一些改变传动比的齿轮组成,用来减速增矩。此系统利用行星齿轮组来完成工作,它靠输入轴和输出轴把动力传递给驱动轮。

计算机利用发动机和变速器上的传感器检测节气门位置(开度)、车速、发动机转速、发动机负荷及制动灯开关位置来精确地控制换挡时间和如何柔和换挡。

自动变速器的另一个优势是:具有自诊模式,可以提前诊断故障所在,并且用仪表盘上的指示灯提示。技术员把测试仪器插入检查孔,收寻故障码、锁定故障位置。

Key to the Exercises

Exercise One
1. 自动变速器
2. 最高挡(直接挡)
3. 行星齿轮装置
4. 超越离合器
5. 倒挡
6. 自诊模块

Exercise Two
1. c　　2. b　　3. a　　4. c

Exercise Three
1. Changing the torque transmitted from the engine crankshaft to the propeller shaft, reversing the vehicle movement and disengaging the engine from the drive line for a long time at parking or coasting.
2. Torque converter, gear system and hydraulic system.

第十一课　液力变矩器

在筑路、起重和交通运输装置中我们常常见到液压装置。

①就像装有手动变速器的汽车一样,装有自动变速器的汽车也需要安装一个装置,使发动机在车轮和变速器齿轮停止转动时能继续运转。

手动变速器的离合器可以使发动机与变速器完全分离,而自动变速器使用的是变矩器。

变矩器是液压耦合器的一种,它不能使发动机与变速器完全独立转动。如果发动机缓慢转动,比如遇见红灯汽车急速,那么传到变矩器的转矩就非常小,因此,保持(停车)仍然需要轻轻地踩制动踏板。

如果汽车停车而你却踩加速踏板,则必须用更大的力踩制动踏板防止汽车行驶。这是因为你踩加速踏板发动机就会加速,就会泵出更多的(变速器)油进入到变矩器,使更大的转矩传递到车轮上。

变矩器由四部分组成：
(1) 泵轮;
(2) 涡轮;
(3) 导轮;
(4) 变速器油。

变矩器壳用螺栓固定在飞轮上,所以它与发动机相同的转速旋转,构成变矩器泵轮的叶片连接在壳上,所以它也与发动机相同的转速旋转。

图1-11-1为液力变矩器构件。变矩器的泵轮是一种离心泵,当它旋转时(变速器)油向外甩,当(变速器)油向外甩时会产生真空,在变矩器中央吸进更多的油。然后,液油会进入到于变速器

相连的涡轮叶片处,涡轮使变速器旋转。涡轮的叶片呈曲线形,这就意味着从(泵轮)外端进入到涡轮的液油在它从涡轮中央出来之前必须改变方向,也就是方向的改变才使涡轮旋转。

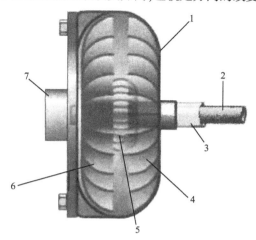

图 1-11-1　液力变矩器各组成构件

1-液力变矩器壳(接飞轮);2-涡轮输出轴(接变速器);3-导轮输出轴(接变速器固定轴);4-泵轮(接变矩器壳);5-导轮;6-涡轮;7-飞轮(接发动机)

　　液油从涡轮中央出来与它进来时运动方向不同,也就是说油液从涡轮出来向泵轮转动相反的方向移动。如果允许液油撞击泵轮,会使发动机减速浪费功率,这就是要用导轮的原因。导轮安装在变矩器正中间,它的作用是涡轮出来的油液撞击泵轮之前改变液油的返回方向,这样神奇地增加了变矩器的效率。

　　导轮是非常厚重的叶片,它几乎可以完全使油液移动方向相反。(导轮中的)单向离合器把导轮与变速器的固定轴相连。因为这种布置,导轮不会随着油液转动,它会与液油相反的方向转动,当油液撞击导轮叶片时迫使油液改变方向。

　　当汽车行驶时会出现一点小麻烦,大约在 40mile/h(64km/h)时,泵轮和涡轮几乎以相同的转速旋转(泵轮应转的更快一点),这时油液从涡轮返回并且以与泵轮运动相同的方向进入泵轮,所以不需要导轮。

　　即使涡轮改变了油液的运动方向并且还是把它向后甩,油液仍然会沿着涡轮旋转方向运动,因为涡轮沿一个方向旋转的速度比油液向后甩的速度更快。如果人站在行驶速度为 60mile/h 的轻型货车(皮卡车)上,以 40mile/h 的速度向后掷球,球仍会以 20mile/h 的速度向前运动,这种现象与涡轮箱类似。油液向后甩,但没有(随涡轮)向前运动速度快。

　　在这种情况下,油液实际撞击导轮叶片的背面,使导轮在单向离合器的作用下成为自由轮,因此不会阻挡油液通过导轮。

　　另外,除使汽车完全停车而不熄火的重要作用外,液力变矩器还会在(怠速)起步时产生更大的转矩。现代液力变矩器可以把发动机扭矩放大 2～3 倍,这种效果只会出现在发动机转速比变速器转速高的情况下。

　　在高速时,变速器逐渐达到发动机的速度,甚至以完全相等的速度转动。理想情况是,变速器与发动机转速完全相同,因为速度差会使功率损失,这就是使用自动变速器比手动变速器的汽车经济性(燃油里程数)差的部分原因。

为了消除这种影响,一些汽车安装了带锁止离合器的变矩器,当变矩器的两根轴达到一定速度时离合器把它们锁在一起,消除滑转增加效率。

Key to the Exercises

Exercise One
1. 用在筑路、起重和交通运输设备中的液压装置
2. 使工作液高速旋转的驱动轮(泵轮)
3. 液压耦合器和液力变矩器
4. 把发动机转矩传给变速器的液压离合器
5. 发动机负荷
6. 与汽车变速器相连的液力变矩器

Exercise Two
1. b   2. c   3. c   4. d

Exercise Three
1. pump, turbine, stator, transmission fluid
2. efficiency of the torque converter

第十二课　差　速　器

动力传动系统中一个非常重要的装置是差速器(图1-12-1),它由主减速器驱动。它安装在两驱动桥之间并且可以使一桥与另一桥以不同的转速转动。当汽车转弯或在不平路面上行驶时两桥转速不同是很必要的。同时,差速器把发动机转矩传递给驱动桥,驱动桥在与主传动轴垂直的转轴上转动。

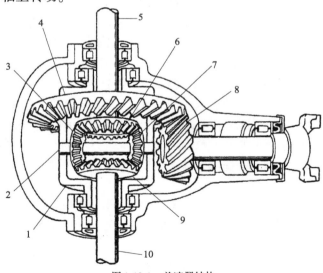

图1-12-1　差速器结构

1-差速器壳;2-行星齿轮轴;3-行星齿轮;4-主减速器环齿(被动环齿);5-半轴;6-半轴齿轮;7-行星齿轮;8-主减速器驱动小齿轮(主动齿轮);9-半轴齿轮;10-半轴

差速器总成利用主传动轴的旋转把发动机的动力传递给驱动桥轴,即使两驱动轴转速不同时差速器也必须有能力把转矩提供给两轴。差速器总成由下列构件组成:差速器壳、主减速器小齿轮、主减速器环齿、差速齿轮和行星小齿轮。

主减速器环齿用螺栓连接在差速器壳一端的凸缘盘上,当主减速器小齿轮驱动环齿时,差速器壳也转动,壳上有安装桥轴的圆孔,在每根轴的一端安装了叫差速齿轮的锥齿轮。实际上,差速齿轮和轴是花键连接,只要差速齿轮转动轴就转动。

差速齿轮间有一定的空间,在这个空间与桥轴垂直方向安装了一根十字轴,这根十字轴叫作行星齿轮轴,十字轴的每一端安装了与两个差速齿轮相啮合的锥齿轮,我们把它们叫作差速器行星齿轮。

汽车的后轮不会总是以相同的转速旋转,当汽车转弯或汽车轮胎直径稍有不同时后轮必须以不同的转速旋转。如果汽车每个桥都和差速器壳之间刚性连接,无论什么时候驾驶员要使汽车转弯车轮都会打滑,发出刺耳的声音,轮胎还会产生磨损。差速器的设计就是为了防止这种问题的产生。

当汽车直线行驶时,环齿、差速器壳、行星齿轮、差速齿轮成一体转动。因为两个行星齿轮对差速齿轮作用相同的力,所以不会绕十字轴转动。因此,差速齿轮与环齿以相同的转速转动,驱动轮也以相同的转速转动。

当汽车转弯时,行星齿轮会在十字轴上转动,这是因为行星齿轮必须绕转动慢的差速齿轮转动,所以,因此行星齿轮要把多余的运动传给外侧的(需要转动)更快的车轮。

Key to the Exercises

Exercise One

1. 主减速器总成
2. 差速器总成
3. 差速齿轮(半轴齿轮)
4. 差速齿轮(半轴齿轮)
5. 行星齿轮
6. 行星齿轮

Exercise Two

1. a   2. c   3. c   4. d

Exercise Three

1. To transmit engine torque (power) to the wheels of the vehicle
2. The differential case, the drive pinion, the ring gear, the differential gears (side gears) and the differential pinions (planetary gears, pinion gears).

第十三课　主 减 速 器

主减速器(图1-13-1)把发动机和变速器的动力传递给车轮驱动车辆。主减速器从传动轴得到动力并把动力方向改变90°使车轮转动。主减速器必须把动力分配给两个驱动轮,当车辆直线行驶时动力均匀分配,这样两个轮子得到相同的动力行驶相同的距离。当汽车转弯时,外边的车轮比里边车轮行驶的距离长,因此,驱动轮必须以不同的驱动力、不同的行驶速度行驶。

如果两个驱动轮以相同的速度行驶,转弯时车轮就会打滑,这样会使轮胎快速磨损。为了避免磨损,主减速器安装了可以在两驱动轮间分配动力的一套齿轮系统,因此,如果必要车轮可以以不同的转速行驶。因为齿轮可以使车轮以不同的转速转动,因此这套齿轮机构称为差速器。许多人把差速器和主减速器混淆,实际上,差速器只是动力分配装置,而主减速器由差速器和其他的装置组成。

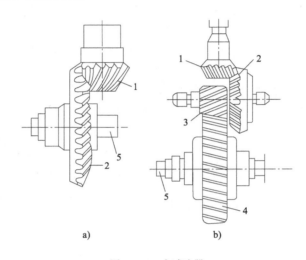

图 1-13-1　主减速器
1-驱动锥齿轮;2-被动锥齿轮;3-主动圆柱齿轮;4-被动圆柱齿轮;5-半轴

不等量的分配动力是很困难的,当一个车轮失去牵引力(冰面上打滑)时甚至会更困难。主减速器总成趋向于将动力传到滑转的驱动轮,而不是传到正常产生驱动力的车轮。多数差速器在这种状况下不能进行动力补偿,这种差速器叫作标准、开式、非防滑差速器。

然而,有一种更复杂的设计结构可以避免这种问题,这种类型的主减速器总成允许两驱动轮适当地打滑。当一个驱动轮比另一个驱动轮转动快时,差速器会感觉到这种变化,差速器会将动力从打滑的驱动轮转移到不打滑的驱动轮上,这种类型的差速器叫作防滑差速器。

主减速器齿轮可以把变速器输出的动力传递给差速器(图1-13-2),主减速器驱动小齿轮、被动环齿和差速器总成通常安装在驱动桥壳内,主减速器通常有四种类型:螺旋齿轮式、行星齿轮式、双曲面齿轮式和链轮式。

驱动小齿轮连接在变速器的输出轴上,被动环齿安装在差速器的壳上,驱动小齿轮和被动环齿可以减速增矩。

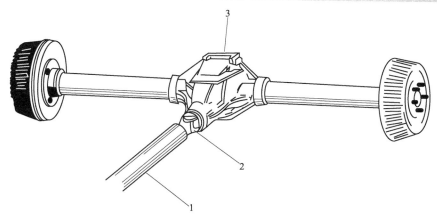

图 1-13-2　主减速器把动力从传动轴传递到驱动轮
1-主传动轴;2-万向节;3-主减速器总成

Key to the Exercises

Exercise One

1. 主减速器
2. 防滑差速器
3. （主减速器）主动轮
4. （主减速器）被动轮
5. 底盘（车架）横梁
6. 底盘（车架）纵梁

Exercise Two

1. b　　2. b　　3. c　　4. d

Exercise Three

1. drive pinion, ring gear, differential and other parts

2. When one drive wheel starts spinning at a much greater speed than the other wheel, the differential "senses" the change. The differential then transfers the power away from the wheel that spins and toward the one that does not spin. This type of differential is called a limited-slip differential.

第十四课　制　动　系　统

汽车运行和控制最重要的因素是制动系统。

为了使行驶的车辆在最短的时间内停车或减速，车辆所具有的动能必须转换成其他形式的能量。减速度受能量转换速度控制。动能是运动的能量，它被转换成热量释放到流过制动系的空气中。

在最短距离内使运行的车辆减速或停车的方式叫制动。制动器是一种摩擦装置，通过摩擦把运动车辆的能量或动能转换成热量。制动系统包括制动器总成，含义是施加制动力。

现代车辆使用两种类型的制动器:车轮制动器和驱动轴制动器。车轮制动器又有三种形式:鼓式制动器、盘式制动器和带式制动器(常用在履带机械上)。

顾名思义,车轮制动器安装在车轮上,图1-14-1所示液压控制鼓式制动器总成。它包括制动鼓、膨胀器、复位弹簧、固定后板、带摩擦衬片的两个制动蹄和支撑销。固定后板或与桥壳的凸缘盘相连或与转向节相连,制动鼓安装在轮毂上,制动鼓内表面与制动蹄衬片之间有一定的间隙。

当踩下制动踏板时,推杆推动制动主缸的活塞,活塞压向制动液,通过油路的制动液进入制动助力器在进入到制动轮缸。液体压力使轮缸活塞向外移动,因此把制动蹄压向制动鼓。松开踏板,活塞回到最初的位置,复位弹簧式制动蹄复位,制动液返回到主缸,制动停止。

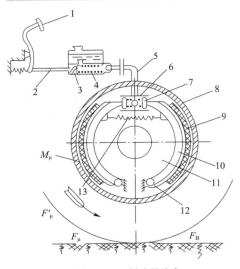

图1-14-1 制动器总成

1-制动踏板;2-推杆;3-制动主缸活塞;4-制动主缸;5-液压油路;6-制动轮缸;7-制动轮缸活塞;8-制动鼓;9-摩擦垫片;10-制动蹄;11-支承板;12-支撑销;13-复位弹簧

驻车制动器在脚制动器失灵时能够停车,因此,它必须和行车制动器分离而且要用钢绳或拉杆(机械式)取代液压系统。

近年来,大多数的汽车都安装防抱制动系统(ABS),最有效的制动压力出现在每个车轮将要抱死之前。当紧急停车猛踩制动器时,会产生刺耳的声音并会在路面上留下橡胶痕迹,制动距离要比正常停车时的长。另外,前轮抱死会失去转向控制,所以无法绕过障碍物;如果后轮抱死,汽车会甩尾。过多打滑会出现另一个问题,那就是轮胎上被烧出许多斑点。

防抱制动系统无论何时检测出车轮抱死,都会通过快速泵压制动器解决这个问题。大多数情况下,只要有车轮抱死就会被泵压,而制动力对其他车轮仍然有效。即使一个或多个车轮在冰面上,这种制动效果可以使你在最短的时间内停车而且还能保持转向控制。

防抱制动系统使用计算机检测每个车轮的转速,当它检测出一个或多个车轮停转或比其他车轮转得慢时,计算机就会发出信号对受影响车轮瞬间解除或施加压力使它们继续转动。这种"泵压"一秒内会达到十次或更多次,远远高于人所能达到的泵压次数。

防抱制动系统包括电子控制单元、液压驱动器和安装在每个车轮上的车速传感器。如果控制单元检测出系统故障,仪表盘上的ABS警报灯就会亮,警告你出现故障。如果有问题出现,防抱制动系统就不会工作而车轮制动器还会正常工作。

Key to the Exercises

Exercise One

1. 驻车制动器
2. 鼓式制动器
3. 盘式制动器
4. 带式制动器

5. 制动系统效率
6. 防抱制动系统

Exercise Two

1. c　　2. c.　　3. d　　4. a

Exercise Three

1. electronic control system, hydraulic actuator, wheel speed sensors
2. To slow down or bring to rest a moving vehicle in the shorted possible distance.

第十五课　转 向 系 统

　　转向系的作用是把驾驶员手中转向盘的旋转运动变成转向轮的偏转运动,通过机械效益或转向助力器放大驾驶员的作用力使转向容易。转向系包括转向器和转向联动装置。转向器的基本类型包括循环球式、齿条—柱塞式和涡轮—蜗杆式。

　　转向为手动或液压助力转向。转向盘的转动作用在联动装置的(各种名称)的杆上,通过杆使转向轮偏转。

　　转向器增加了从转向盘到转向垂臂的力,使转向轮转向容易。①一些车辆上还安装了液压助力转向系统,目的是减小驾驶员转动转向盘的作用力,并且减小道路传给转向盘的振动。助力器与转向器成一体(图 1-15-1)。转向器包括壳、蜗杆、球螺母、与柱塞成一体的齿条、液压助力器及与转向垂臂成一体的齿扇。

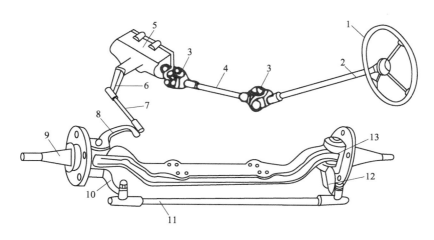

图 1-15-1　转向系

1-转向盘;2-转向杆;3-万向节;4-转向器轴;5-转向器;6-转向垂臂;7-转向纵拉杆;8-转向节臂;9-左转向节;10、12-转向臂;11-转向横拉杆;13-右转向节

　　球螺母刚性地安装在柱塞齿条的内部。为了减小蜗杆与螺母之间的摩擦,循环球在螺纹之间循环(图 1-15-2)。齿条柱塞上安装有弹簧垫片确保与转向器壳紧固。由于螺母沿着螺杆的位移,把转向轴的旋转运动转换成齿条柱塞的直线运动。齿条柱塞的齿使齿扇转动,因此也就使与垂臂连在一起的转向轴转动。

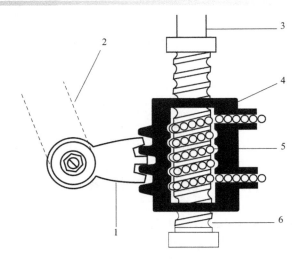

图 1-15-2 循环球式转向器部件
1-齿扇;2-转向垂臂;3-转向轴;4-球螺母齿条;5-循环球轴承;6-蜗杆

车辆转弯时作用在转向器上的力应该传递到转向轮上,这个工作由转向联动装置完成,联动装置由转向垂臂、横拉杆、纵拉杆、转向臂和左、右转向节臂组成。转向联动装置的设计是为了防止所有车轮在转弯时侧滑,确保转弯容易且轮胎磨损程度最小。为了达到这个目的,所有车轮的几何轴线必须相交在一点,即所有车轮运动的圆形轨迹的共同中心。

液压助力器的双作用(双腔式)叶片泵、油箱及滤清器都安装在发动机上,由与发动机曲轴相连的三角皮带或齿轮驱动,泵与控制阀通过两根油管相连:从泵输油的输油管及使油返回到油箱的回油管。

Key to the Exercises

Exercise One
1. 方向盘
2. 转向器
3. 转向系的灵敏性
4. 转向轮
5. 转向半径
6. 转向节

Exercise Two
1. b   2. a   3. b   4. b

Exercise Three

1. To convert the rotary movement of the steering wheel in the driver's hands into angular turn of the front wheels, and to multiply the driver's effort by leverage or mechanical advantage or booster so as to make it fairly easy to turn the wheels.

2. Increasing the force transmitted from the steering wheel to the pitman arm and thus facilitating the turning of the steerable wheels.

第十六课　车架和悬架系统

车架是承载梁,由低碳钢制成,用焊接、铆接或螺栓连接在一起。它是车身与许多系统的支撑结构,把载荷通过悬挂系统和车桥传递给车轮。在允许变形范围内,它承受静态和动态载荷。它具有一定的硬度和强度,在车辆运行过程中抵抗它所承受的扭力和弯力。

车架(梁的截面)是箱型、管型、槽型和角型结构。横梁增加了车架的强度,用来支撑发动机和车轮。

悬架系统(图1-16-1)的功用是吸收道路振动并防止振动传递到车辆的其他部件上。它保护部件免受冲击和动载的作用,在这种情况下,其他部件在车辆运行过程中变得更安全,而且还增加了它的使用寿命。由于悬架系统的提供使乘车人不会受震动的痛苦,使他们的旅行更舒适不会疲劳。悬架系统还能使车辆在行驶过程中遇到冲击保持稳定。

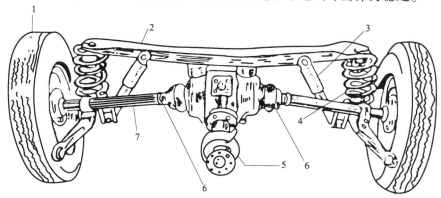

图1-16-1　后轮独立悬架系统
1-轮胎;2-横梁;3-减振器;4-螺旋弹簧;5-中央轴承;6-万向节;7-半轴

弹簧、扭杆和用来起到连接作用的部件都叫悬架系统部件。

钢板(叶片)弹簧由片状弹簧钢制成,由许多片叠加在一起,它们靠夹子或中央螺栓连接在一起。每一片的长度是递减的,目的是弹簧总成可以作为一个挠性梁并且具有均匀的强度,最长的那一片叫主钢板。

螺旋弹簧由特殊的螺旋钢丝制成,弹簧的截面是圆形的并且具有一定直径来达到所需要的硬度。钢丝绕成螺旋的形状。弹簧在高温下形成、冷却并且要经过热处理达到所需要的弹性。当车轮在路面上遇到振动时,弹簧压缩吸收振动能量。

扭杆通过抵抗扭转起到弹簧的作用。螺旋弹簧靠压缩吸收作用力,钢板弹簧靠弯曲吸收作用力,而扭杆靠扭转吸收作用力。扭杆由钢制成。在某些车辆上扭杆横向安装,而在另一些车辆上它沿着车架纵向安装。扭杆的一端固定在车架上不能转动,另外一端承受转矩,这种安装方式使杆可以在自由端扭转。在扭转(自由)端的支撑臂用来安装车轮。扭杆也有弹性比,弹性比取决于杆的直径和长度。

对于悬架系统来说,单独使用弹簧不会令人满意,汽车悬架系统理想的弹簧应该是快速吸收道路振动慢慢返回到初始位置。对于弹簧来说这是不可能做到的,因为它的属性是振荡。弹簧既要有一定的柔性还要有一定的硬度。①太软的弹簧会产生太大的位移,太硬的弹簧会使舒适性下降。把软弹簧与减振器结合起来使用可以获得满意的舒适程度(平顺性)。

目前汽车最常见的减振器是液压减振器,工作原理是:液体不可压缩,液体经过小孔时受到的阻力迫使弹簧受到一个拉力,快速使弹簧的振荡衰减。

Key to the Exercises

Exercise One

1. 叶片(钢板)弹簧
2. 螺旋弹簧
3. 扭杆
4. 底盘
5. 气体弹簧
6. 悬挂系统力学

Exercise Two

1. b   2. c   3. b   4. c

Exercise Three

1. the leaf spring, the coil spring, the torsion bar
2. to support all of parts to hold them together and withstand shocks, vibrations

>>> Part 2

Engineering and Construction Machineries

◎ Unit 1 Texts

◎ Unit 2 Reference Translations and Answers

Unit 1   Texts

Lesson 1

Bulldozers

Bulldozer is a earth moving machine, which consists of tractor and a blade mounted on the crawler and wheel tractors(Fig. 2-1-1). ①In laying a road bed bulldozers are employed to strip off the vegetation covering, remove bushes, stumps and small trees, dig and transport soil from cuts to a pile, build fills, push soil from cuts to spoil banks, move soil when shaping earth bed on hillside, grade and level soil piled up by earth diggers, fill up trenches, ditches and pits, transport dirt to bridge abutments.

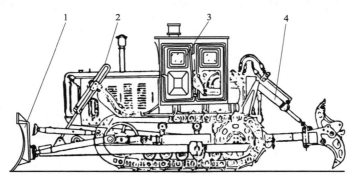

Fig. 2-1-1   Crawler bulldozer
1-blade; 2-hydraulic system; 3-crawler tractor; 4-hydraulic ripper

A bulldozer can likewise be used auxiliary jobs such as: cleaning the path for scrapers or dump trucks, helping scrapers forward when their bowls are being filled, service with power shovel or hydro-mechanical equipment.

In winter-time bulldozers are employed to remove snow from motor roads, airfields and municipal squares.

At storage depots and construction sites bulldozers transport, pile and load crushed stone, sand, gravel and other loose materials, develop small gravel and sand quarries and supply stone material to the feeding devices of mixers.

The use of bulldozers for moving soil and loose construction materials are highly expedient at distance less than 100 meters. Bulldozers can erect fills and dig cut to a height or to a depth of up to 3 meters.

Bulldozers can be classified as follows:

(1) By the mounting of blade: straight bulldozer, whose blade is perpendicular to the longitudinal axis of the tractor (with the blade in this position the bulldozer pushes cut earth forward); angle or universal bulldozer, where the blade can be set either perpendicularly to the longitudinal axis of the machine or at angle (usually 60°) to cast dirt to either side. A universal blade can also be tilted in a vertical plane by an angle of ±5° to 8°. A universal blade can push the earth forward and aside, fill up ditches and trenches and shape an earth bed;

(2) By the control system: cable bulldozer, whose blade is raised and lowered by a cable hoist driven by a winch mounted on the tractor; hydraulic bulldozer, whose blade is operated by hydraulic cylinders supplied with work fluid by a pump under pressure.

With hydraulic bulldozers there is a force penetration of the blade. hydraulic bulldozers are gaining an ever broader recognition in the world.

(3) By the type of carrying vehicle: there are crawler and wheel tractors and special air-tyred tractors.

The advantages of crawler-bulldozer are the following:

(1) Ability to deliver greater traction effort, especially in operation on soft footing, such as loose or muddy soil;

(2) Ability to travel over muddy soil;

(3) Ability to operate in rocky formations where rubber tyres might be seriously damaged;

(4) Ability to travel over rough surfaces, which may reduce the cost maintaining haul roads;

(5) Greater floatation because of the lower pressure under the tracks;

(6) Greater use versatility on jobs.

The advantages of wheel-mounted bulldozer are the following:

(1) Higher travel speed on the jobs from one job to another;

(2) Elimination of hauling equipment to transport the bulldozer to a job;

(3) Greater output, especially when considerable traveling is necessary;

(4) Less operator fatigue;

(5) Ability to travel on paved highways without damaging the surfaces.

New Words

1. bulldozer ['buldəuzə]	n.	推土机
2. pendant [ˌpendənt]	n.	悬挂,下垂物;垂饰
	adj.	悬挂的
3. attachment [ə'tætʃm(ə)nt]	n.	附件;附属装置,连接物
4. strip [strip]	vt.; vi.	剥,磨掉
	n.	条;带

5. vegetation [ˌvedʒɪˈteɪʃ(ə)n]　　　n.　　　植被
6. bush [buʃ]　　　n.　　　灌木,衬套;轴瓦
7. stump [stʌmp]　　　n.　　　根株,残余
8. fill [fil]　　　n.　　　填方,填土,路堤
　　　　　　　vt.;vi.　　　填满,充满
9. spoil [spɔil]　　　vt.　　　损坏
　　　　　　　n.　　　赃物
10. hillside [ˈhilsaid]　　　n.　　　山腰
11. trench [tren(t)ʃ]　　　n.　　　沟;堑壕
　　　　　　　vt.;vi.　　　掘沟
12. ditch [ditʃ]　　　n.　　　沟,渠
　　　　　　　vt.;vi.　　　筑沟
13. digger [ˈdigə]　　　n.　　　挖掘机
14. pit [pit]　　　n.　　　坑,穴
　　　　　　　vt.;vi.　　　弄凹
15. abutment [əˈbʌtm(ə)nt]　　　n.　　　桥台,拱座,支承
16. likewise [ˈlaikwaiz]　　　adv.　　　同样地
　　　　　　　conj.　　　也,而且
17. scraper [ˈskreipə]　　　n.　　　铲运机
18. dump [dʌmp]　　　n.　　　垃圾,翻斗倾卸车
　　　　　　　vt.　　　倾卸
19. bowl [bəul]　　　n.　　　斗,碗
20. hydro-mechanical [ˌhaidrəumiˈkænikəl]　　　adj.　　　流体力学的
21. airfield [ˈeəfiːld]　　　n.　　　飞机场
22. municipal [mjuːˈnisip(ə)l]　　　adj.　　　市政的,地方的
23. depot [ˈdepəu]　　　n.　　　仓库
24. crush [krʌʃ]　　　vt.;vi.;n.　　　压碎
25. quarry [ˈkwɔri]　　　n.　　　石场,石矿,采石
26. expedient [ikˈspiːdiənt; ek-]　　　adj.　　　合适的,有利的
27. erect [iˈrekt]　　　vt.　　　树立;建立
　　　　　　　adj.　　　直立的
28. mixer [ˈmiksə]　　　n.　　　混合机;搅拌机
29. winch [wintʃ]　　　n.　　　绞盘;绞车
30. penetration [peniˈtreiʃ(ə)n]　　　n.　　　入土,贯穿,渗透
31. air-tyred [eətaiəd]　　　adj.　　　轮胎的,气胎的

Phrases and Expressions

1. dump truck　　　自动卸货车

Part 2 Engineering and Construction Machineries

2. hydro-mechanical equipment　　　　　　液压装置,液压设备
3. construction site　　　　　　　　　　　　建筑工地
4. feeding device　　　　　　　　　　　　　进料装置
5. earth bed　　　　　　　　　　　　　　　地基

Notes

①In laying a road bed bulldozers are employed to strip off the vegetation covering,... transport dirt to bridge abutments.

在铺设路基中,推土机用来剥离草皮覆盖层、清除灌木、树桩和小树,挖土并把土从挖土处运到某一堆土处,筑路堤、推土至弃土堆、在山坡上筑路基时推土、平整挖土机堆积的土堆、填平沟、渠和坑洼、运淤泥到桥墩处。

Exercise One

Put the following expressions into Chinese:
1. a pendant attachment mounted on crawler and wheel tractors
2. motor roads, airfields and municipal squares
3. the ability to travel over muddy soil
4. bridge abutment
5. earth bed
6. hydraulic bulldozer

Exercise Two

Translate the following sentences into Chinese.
1. The blade of a straight bulldozer is perpendicularly to the longitudinal axis of the tractor.
2. Hydraulic bulldozers are gaining ever broader recognition in world practice.
3. Hydraulic bulldozer blades are approaching the same speed as cable control blades.
4. The crawler mounted bulldozer is usually superior to the wheel-mounted.

Lesson 2

Scrapers

A scraper removes soil slice by slice, transports and places it in an earthen structure, or pushes it to a dump and then levels it. ①When moving over freshly dug soil the scraper partially compacts it.

Scrapers are used extensively in road-making, industrial and hydro-technical construction, in baring jobs in quarries and in earth working jobs.

Scrapers can operate on a wide variety of soils, including black earth, sand and clay. On moist clay and black earth scrapers are not very efficient as the soil sticks to the walls of the bowl and clogs it; they are quite useless on boggy soil as the wheels sink into the mud.

Loose sand does not fill the bowl well and leaves it with difficulty. Scrapers prove most useful on moderately wet sand soils and loams as they fill the entire volume of bowl. Scrapers must not be used on soils containing large stones. Very heavy soils should be ripped beforehand. Work is underway at present in countries on paddle feeders and elevators, which help to fill the bowl better, load the scraper more uniformly, and utilize the traction power of the tractor more effectively. Scrapers can be classified as follows:

(1) By the geometrical capacity of the bowl:

1.5, 3.0, 6.0, 10.0, 15.0, 25.0m^3, etc.

(2) By the mode of locomotion:

trailer, semi-trailer and self-powered.

(3) By the method of discharge:

free, half forced and forced discharge.

(4) By the system of control:

When a scraper has to operate on heavy soil the pull of one tractor or a two-wheel air-tired tractor may be insufficient to cut a slice and fill the bowl. In such cases the leader is helped forward by a crawler or a four-wheel air-tired pusher tractor. Scrapers are the result of a compromise between best loading and best hauling machines, as must be expected of any composite machine, they are not superior to other equipment in both loading and hauling. Power shovels, draglines, and belt loaders usually will surpass them in loading only, while trucks may surpass them in hauling only, especially when long, well-maintained haul roads are used. There are two types of scrapers, based on the type of tractor used to pull them:

the crawler-tractor-pulled and the wheel-tractor-pulled. The pusher tractor presses against the rear bumper of the scraper and together with the towing tractor produces the force necessary to fill the bowl. In view of the fact that the process of cutting the soil and filling the bowl takes a relatively short period of the scraper working cycle, one pusher tractor can be used to service several scrapers.

[2]For relatively short haul distance the crawler-type tractor, pulling a rubber tired self-loading scraper, can move earth economically. The high drawbar pull in loading a scraper, combined with good traction, even on poor haul roads gives the crawler tractor an advantage for short hauls. However, as the haul distance is increased, the low speed of a crawler tractor is a disadvantage compared with a wheel tractor.

Unless the loading operation is difficult, a crawler tractor can load a scraper without the aid of a bulldozer. However, if there are several scrapers united on a job, the increased output resulting from using a bulldozer to help load the scrapers usually will justify the use of a bulldozer.

For longer distance the higher speed of a wheel-type tractor-pulled self-loading scraper will permit to move earth more economically than a crawler-type tractor. Although the wheel-type tractor can not deliver as great as traction effort in loading a scraper, the higher speed will offset the disadvantage in loading when the haul distance is sufficiently long.

A scraper is loaded by lowering the front end of the bowl until the cutting edge, which is attached to end extending across the width of the bowl, enters the ground and, at the sametime, raising the front apron to provide an open slot through which the earth may flow into the bowl. As the

Part 2　Engineering and Construction Machineries

scraper is pulled forward, a strip of earth is forced into the bowl. This operation is continued until the bowl is filled or until no more earth may be forced in. The cutting edge is raised and the apron is lowered to prevent spillage during the haul trip.

The dumping operating consists of lowering the cutting edge to the desired height above the fill, raising the apron, and forcing the earth out between the blade and the apron by means of a movable ejector mounted at the rear of the bowl. The operation of scraper is shown in Fig. 2-2-1.

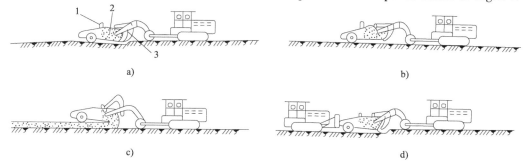

Fig. 2-2-1　Operation of scraper
a) Loading; b) transportation; c); discharge; d) Loading helped by bulldozer
1-discharge plate; 2-bowl; 3-bowl gate

New Words

1. scraper['skreipə]	n.	铲运机,刮刀;刮板,刮工
2. slice[slais]	n.	片,一部分
3. compact[kəm'pækt]	adj.	紧密的,紧凑的
	n.	小型汽车,契约,合同
	vt.;vi.	压缩,压实
4. extensively[iks'tensivli]	adv.	广泛地,多方面地
5. clay[klei]	n.	黏土
6. moist[mɔist]	adj.	潮湿的
7. clog[klɔg]	vt. n.	阻塞
8. boggy['bɔgi]	adj.	沼泽的
9. mud[mʌd]	n.	泥泞,泥浆
10. moderately['mɔd(ə)rətli]	adv.	适中,适宜
11. sandy['sændi]	adj.	沙的,含沙的
12. loam[ləum]	n.	壤土;卢姆土
13. rip[rip]	vt.	裂开;撕开,破裂
14. beforehand[bi'fɔːhænd]	adv.	事先
15. pusher['puʃə]	n.	后推机车,后推者
16. bumper['bʌmpə]	n.	保险杠,缓冲器
17. underway['ʌndə'wei]	adv.	正在发展,开始运动

18. paddle[ˈpæd(ə)l]	n.		刮板,桨,踏板
	vt. ;vi.		涉水,划桨
19. haul[hɔːl]	vt. ;vi.		拖,拉
20. drawbar[ˈdrɔːbɑː]	n.		拉杆,牵引杆
21. justify[ˈdʒʌstifai]	vt.		证明……是正确的;认为……有理
22. exceed[ikˈsiːd; ek-]	vt. ;vi.		大于;多于,超越
23. offset[ɔfˈset]	vt.		抵消,弥补
24. breakeven[ˌbreikˈiːvən]	adj.		临界的,保本的
25. analysis[əˈnælisis] (analyses)	n.		分析,分解

Phrases and Expressions

1. slice by slice   一小片一小片地;一层一层地
2. earthen structure   土方结构
3. hydro-technical construction   土工建筑
4. two wheel air-tired tractor   两轮充气轮胎式牵引车(拖拉机)
5. in view of the fact that   鉴于……这一事实

Notes

①When moving over freshly dug soil. . . .

当驶过新挖的土壤时……

现在分词短语作时间状语,前面常有when 或while,它表示的动作与谓语表示的动作同时进行(也可认为 when 后的 it is 被省略)。

②For relatively short haul distance the crawler-type tractor,pulling a rubber-tired self-loading scraper,can move earth economically.

运距较短时,履带牵引车助推轮胎、自行装载式铲运机可以更经济的运土。

pulling a rubber-tired self-loading scraper   现在分词短语作定语,用逗号隔开,具有较强的状语意义。

Exercise One

Put the following expressions into Chinese：

1. best loading and best hauling
2. well-maintained haul road
3. crawler tractor
4. earthen structure
5. rubber tired self-loading scraper
6. road-making,industrial and hydro-technical construction

Exercise Two

Translate the following sentences into Chinese.

1. Scrapers towed by crawler tractors have a very low traveling speed.

2. As the haul distance is increased, the low speed of crawler tractor is a disadvantage compared with a wheel tractor.

3. Scrapers prove most useful on moderately wet sand soils and loams as they fill the entire volume of bowl.

4. As the scraper is pulled forward, a strip of earth is forced into the bowl.

Lesson 3

Loaders

The mechanization of loading and unloading operations is of paramount importance for the comprehensive mechanization construction. In this connection, various kinds of loaders are used, classified as follows:

by their purpose: for handling piece or loose loads;

by the character of working process: for intermittent-operation-single-bucket, for continuous-operation-multi-bucket;

by the running gear: crawler or air-tyred;

by the kind of engine: electric motor powered from mains, internal-combustion engines or diesel-electric drives.

Unloading equipment handles materials delivered on conventional railway cars and is available in modifications to suit the nature of load.

Special unloading devices are used to handle sand, gravel and cement. Other kinds of material are unloaded with the aid of universal hoisting machinery (grab-bucket cranes, etc.).

In terms of how they operate single bucket loaders can be subdivided into front loaders (the bucket is tilted forward). Swing (part-revolving) loaders (the bucket is swung sideways to be dump), rear loaders and combination types.

[1]Front and swing loaders are provided, as a rule, with changeable equipment, such as large and small buckets, grabs, forks, hooks etc., then its application is large expanded. These loaders are used for the loading and unloading of loose materials, such as earth, pebbles, coal, sand, etc.. It may also be used in operation like pile demolishing in mines. As an auxiliary equipment is set up, it may be used in lieu of a bulldozer in road-paving, cleaning up working face, leveling earth deposits, etc.. This loader is an ideal equipment for use in miners, road building, harbour, wharf and forest region operations, as well as other basis construction work.

In modern times articulated frame loaders are widely used. These loaders greatly improve the performance of steering. They have small turning radius, high mobility and agility, if it is not strict alignment necessary when loading or unloading. Satisfactory effect can be attained even when working in a restricted area.

In its transmission system, hydraulic torque converter is applied. It makes loader high efficiency and afford protection to the working parts of the engine and transmission system.

Operating equipment of a single bucket loader consists of lift arms, lift cylinders, tilt cylinders, tilt levers, tilt links and a bucket as shown in Fig. 2-3-1. There are two lift cylinders that makes two arms with bucket lift or lower. Two tilt cylinders extend or retract to turn the bucket, thus material is loaded into bucket or unloaded from it.

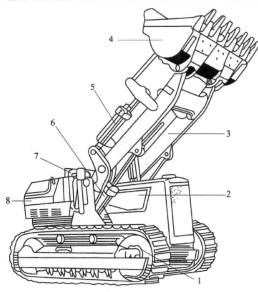

Fig. 2-3-1 Crawler load mechanisms
1-crawler-running gear; 2-engine; 3-lift arm; 4-bucket; 5-tilt cylinder; 6-lift cylinder; 7-cab; 8-fuel tank

Presently front loaders with positive displacement hydraulic drives for the attachment are most widely used. The output of multi-bucket loaders is 40 to 60 percent higher than that of single-bucket machines at the same power rating, it is expedient to use them in brickyards, prefabricated concrete products plants, railway stations with large volumes of loading-unloading operations and also for loading loose materials. In addition, they are suited for size grading of loose materials, for which purpose they are provided with special vibrating screens.

Multi-bucket loaders are very efficient in unloading flatcars with side dumping of the handle material. These loaders can be applied in the production lines of prefabricated construction products plants and also in road building. In the latter case they are used for loading sand and gravel into drying drums and mixers.

The working member of the loader is a screw feeder consisting of two right-hand and left-hand screws, which are arranged on both sides of a bucket elevator. When the feeder rotates, the loaded material is delivered to the buckets to make the scooping easier. A scraper is secured underneath the screw feeder. The elevator of the multi-bucket loader usually discharges the material onto belt conveyors that deliver it to transport facilities.

Some loaders discharge the handled material into transport facilities through hoppers or chutes.

New Words

1. paramount [ˈpærəmaunt]　　　　　*adj.*　　　最高的，至上的
2. comprehensive [kɔmpriˈhensiv]　　*adj.*　　　综合的，广泛的，有理解
3. intermittent [IntəˈmIt(ə)nt]　　　 *adj.*　　　间歇的，间断的
4. grab [græb]　　　　　　　　　　*vt.; vi.*　　抓取，强夺
　　　　　　　　　　　　　　　　　n.　　　　抓取，强夺，抓斗
5. pebble [ˈpeb(ə)l]　　　　　　　　*n.*　　　　卵石，小石子

Part 2 Engineering and Construction Machineries

6. demolish [di'mɔliʃ]　　　　　　vt.　　　　排除,推翻
7. deposit [di'pɔzit]　　　　　　　vt.　　　　存放,沉积
　　　　　　　　　　　　　　　　n.　　　　存款,沉积
8. wharf [wɔːf]　　　　　　　　　　n.　　　　码头
　　(wharves;wharfs)
9. articulate [ɑː'tikjuleit]　　　　vt.;vi.　　咬合,连接
10. mobility [mɔ'biliti]　　　　　　n.　　　　机动性,流动性,易变性
11. agility [ə'dʒiləti]　　　　　　　n.　　　　敏捷,活泼
12. prefabricate [priː'fæbrikeit]　　v.　　　　预制
13. chute [ʃuːt]　　　　　　　　　　n.　　　　漆布,斜道,溜槽

Phrases and Expressions

1. in the connection　　　　　　　　　　　　在这方面;就此而论
2. intermittent-operation-single-bucket　　　单斗间歇操作
3. in lieu of　　　　　　　　　　　　　　　代替
4. grab-bucket　　　　　　　　　　　　　　抓斗
5. continuous -operation- multi- bucket　　多斗连续操作
6. prefabricated concrete product　　　　　混凝土预制件

Notes

①Front and swing loaders are provided, as a rule, with changeable equipment, such as large and small bucket, grabs, forks, hooks, etc.
通常,前卸式和侧卸式装载机都可配备可替换的设备,如:大、小料斗,抓斗,货叉,吊钩等。as a rule 是插入语。

Exercise One

Put the following expressions into Chinese:
1. mechanization of loading and unloading operations
2. articulated frame loaders
3. prefabricated concrete products plants
4. multi-bucket loader
5. intermittent operation single bucket loader
6. torque converter

Exercise Two

Translate the following sentences into Chinese.
1. These loaders can be applied in the production lines of prefabricated construction products plants and also in road building.
2. Practice has shown that excavators are less effective in the capacity of load in quarries and storage of materials than loaders.

3. Crawler loaders have a high passability and develop a great thrust effort.

4. Some loaders discharge the handled material into transport facilities through hoppers or chutes.

Lesson 4

Excavators

①A power shovel is an earth digging machine with an operating member in the form of a bucket which digs soil, carries it a small distance and dumps it in a pile or onto transporting facilities.

Revolving shovels are employed in all branches of construction, to drain and irrigate land, extract useful minerals, in mining for baring operations and the open cast excavation of coal and ore. Revolving shovels can be classified by their purpose, bucket capacity, degree of universality, type of power plant, running gear, control system.

Depending on the bucket capacity and purpose, revolving shovels can be placed in the following groups:

(1) Building shovels with bucket capacities of $0.15\text{-}4m^3$, used together with various alternative attachments for construction and auxiliary jobs and in small quarries for excavating building materials.

(2) Quarry shovels with bucket capacities of $4\text{-}8m^3$, intended for extracting raw materials in quarries, handling rock after blasting operations and digging very heavy soils.

(3) Stripping shovels with capacities of $4\text{-}10m^3$, employed to remove the capping and dump it in piles in open-cast mining.

(4) Walking dragline with bucket capacities from 4 to $50m^3$ for digging deep cuts in soft soil in hydro-technical construction.

Revolving shovels are designed to operate with various alternative attachments (also called rigs or fronts): forward shovel (Fig. 2-4-1a), pull shovel (backhoe) (Fig. 2-4-1b), dragline (Fig. 2-4-1c), clamshell (Fig. 2-4-1d), jib crane (Fig. 2-4-1e), tower crane, pile driver (Fig. 2-4-1f) and frozen soil ripper. Other equipment includes knife plane, ditch filler, weight-dropping tamper, etc..

A revolving shovel is composed of the following main parts: front attachment (bucket, boom, stick and hydraulic cylinder); turntable carrying a power plant, with mechanisms and operating equipment; travel unit with crawlers or wheels supporting the main frame.

Operating cycle of a dipper shovel (Fig. 2-4-2): digging, raising the bucket up from the face base B; bringing the stick forward C; pulling the bucket away from the face D; swinging the boom to the dumping place E; adjusting the bucket for dumping F; dumping the bucket by opening its door and turning it round back to the face and setting it at the same time in its initial position A.

The power plant of a shovel may be a diesel engine, an electric drive powered from an external supply, or a diesel-electric drive.

Part 2 Engineering and Construction Machineries

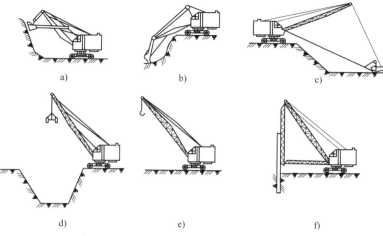

Fig. 2-4-1 Front attachments of revolving shovels

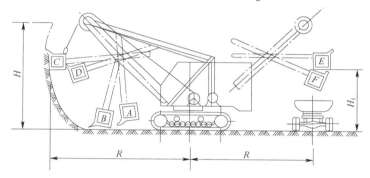

Fig. 2-4-2 Operating cycles of revolving shovels

Excavators fitted with a hydraulic drive for rigs are referred to as excavators with a rigidly suspended working member, as distinguished from excavators with flexible suspension of the working member.

The summary force appearing on the cutting lip of the bucket in the process of digging is usually resolved into a tangential and parallel digging forces. In an excavator with rigid suspension of the bucket the normal digging forces which usually tend to raise the bucket from the soil are overcome by the full weight of the machine, while in an excavator with flexible suspension, only by the weight of the working attachment. This feature of the rigid suspension permits an increase in the output, especially when digging at considerable depths.

Excavators with rigid suspension, feature a number of important design, technical and service advantages over excavators with flexible suspension.

(1) Design advantages:

①Stepless control of traveling speed and traction force permitting simple reversing.

②The possibility of using a multi-motor drive without increasing the total mass of the machine and making it unnecessary to install transmissions requiring attentive attendance.

(2) Service advantages:

Comparatively simple control of the operating conditions of the actuated working mechanisms and working members (speeds). Simple conversion of rotary motion into reciprocating.

(3) Technical advantages:

The possibility of digging only by turning the bucket, with the dipper stick being stationary with respect to the boom, which permits excavation in crowded localities. With a bucket of the same capacity the overall dimensions of a hydraulic excavator are considerably less than those of an excavator with flexible suspension.

(4) Economical advantages:

①Due to wide range of changeable attachments, excavators with rigid suspension permit a reduction in the volume of manual labor and increase productivity.

②It is easy to gang up these machines from standard units, which simplifies their manufacture and repair.

Excavators with rigid suspension can conveniently be rendered automatic, improving, thereby, the working conditions for the machine operator and raising the time utilization factor.

In single-bucket excavators with rigid suspension the working members, running gear, slewing mechanism and other units are actuated by means of hydraulic drives and hydraulic motors.

New Words

1. facility [fə'siləti] n. (常用复数)设备,设施
2. intermittent [ˌintə'mit(ə)nt] adj. 间歇的,断续的
3. ditcher ['ditʃə] n. 挖沟机,挖沟者
4. simultaneously [ˌsiməl'teiniəsli] adv. 同时进行的
5. trench [trentʃ] n. 沟,堑
   vt. 挖沟,挖槽
6. irrigate ['irigeit] vt.;n. 灌溉
7. mineral ['min(ə)r(ə)l] n. 矿物
8. mining ['mainiŋ] n. 采矿,矿业
9. bare [beə] vt. 揭露,剥离
   adj. 裸露的,空的
10. open-cast ['əup(ə)n-kɑːst] adj. 露天开采的
    n. 露天矿
11. ore [ɔː] n. 矿物,矿石
12. bucket ['bʌkit] n. 铲斗
13. blasting ['blɑːstiŋ] n. 爆破
14. capping ['kæpiŋ] n. 盖岩层,剥离物
15. dragline ['dræglain] n. 拉铲挖掘机
16. semi-universal ['semijuːni'vəːs(ə)l] adj. 半通用的
17. rig [rig] vt.;n. 工作装置;装备

Part 2 Engineering and Construction Machineries

18.	backhoe [ˈbækhəu]	n.	反铲
19.	jib [dʒib]	n.	悬臂,吊机臂
20.	clamshell [ˈklæmʃel]	n.	抓斗,合瓣式挖掘机
21.	ripper [ˈripə]	n.	松土机,耙路机
22.	tamper [ˈtæmpə]	n.	夯,打夯机
23.	widespread [ˈwaidspred]	adj.	普遍的,分布广的
24.	replaceable [riˈpleisəbl]	adj.	可替换的
25.	main [mein]	n.	电源
26.	actuate [ˈæktjueit]	vt.	开动,驱动
27.	widen [ˈwaidn]	vt.	加宽
28.	caterpillar [ˈkætəpilə]	n.	履带
29.	essentially [iˈsenʃ(ə)li]	adv.	实质上
30.	expressly [iksˈpresli]	adv.	明显地
31.	pontoon [pɔnˈtuːn]	n.	平底船,起重机船
32.	abandon [əˈbænd(ə)n]	vt.	废弃
33.	boom [buːm]	n.	起重臂,吊杆
34.	initial [iˈniʃəl]	adj.	最初的,开头的

Phrases and Expressions

1. revolving shovel   回转式挖掘机
2. building shovel   组合式挖掘机
3. quarry shovel   采石挖掘机
4. stripping shovel   剥离式挖掘机
5. walking dragline   移动式拉铲挖掘机
6. universal shovel   通用式挖掘机
7. dipper shovel   单斗挖掘机
8. running gear   行走装置
9. tower crane   塔式起重机
10. weight-dropping tamper   落锤式的夯
11. pile driver   打桩机

Notes

①A power shovel is an earth-digging machine with an operating member in the form of a bucket which digs soil, carries it a small distance and dumps it in a pile or onto transporting facilities.

Exercise One

Put the following expressions into Chinese:

1. attachments of hydraulic excavator

2. the structure, hydraulic system and digging geometry of the excavator

3. the effort of weight and the force of inertia

4. the boom of dragline

5. working capabilities of dipper shovel

6. revolving shovel

Exercise Two

Translate the following sentences into Chinese.

1. The shovels with bucket capacities of 4-10m^3, are employed to remove the capping and dump it in piles in the open-cast mining.

2. This feature of the rigid suspension permits an increase in the output, especially when digging at considerable depths.

3. A revolving shovel is composed of front attachment, turn table and travel unit.

4. Excavators with rigid suspension can conveniently be rendered automatic, improving, thereby, the working conditions for the machine operator and raising the time utilization factor.

Lesson 5

Graders

Self-powered graders are used in road construction to excavate side ditches, shape the surfaces and sides of fills and cuts and give them the required gradients.

Powerful self-propelled graders can also be utilized for laying earth beds at zero elevation, for leveling and auxiliary jobs, and for building platforms and making profile cuts and banks. In wintertime, self-powered graders are used to clean roads of compacted snow.

A self-powered grader is a highly manoeuverable machine, and its blade can be set at various angles both horizontally and vertically or brought out sideways.

According to the wheel numbers, self-powered graders can be divided into four-wheel type and six-wheel type. Expression is logarithms of steerable wheels × logarithms of driving wheels × total logarithms of wheels, such as $1 \times 1 \times 2; 2 \times 2 \times 2; 1 \times 2 \times 3; 1 \times 3 \times 3; 3 \times 2 \times 3; 3 \times 3 \times 3$. The more the steerable wheels, the smaller the turning radii are. The more the driving wheels, the more the tractions are.

The Grader Type F155 is a construction machine, shown as Fig. 2-5-1.

The engine is a 5-cylinder 4-stroke diesel engine with direct injection and is air cooled. The cylinders are mounted in line. The drive comes from the engine to torque converter mounted on the engine, 4-gear power shift gear box with transfer box, through a prop shaft to the rear axle, then to the wheels via tandem chains.

Use of the hydrostatic front wheel steering combined with chassis articulation gives the grader an exceptionally high manoeuverability with a minimum turning radius. Because of the high-powered and the versatility of the working equipment the grader F155 can be operated smoothly and economically.

Part 2　Engineering and Construction Machineries

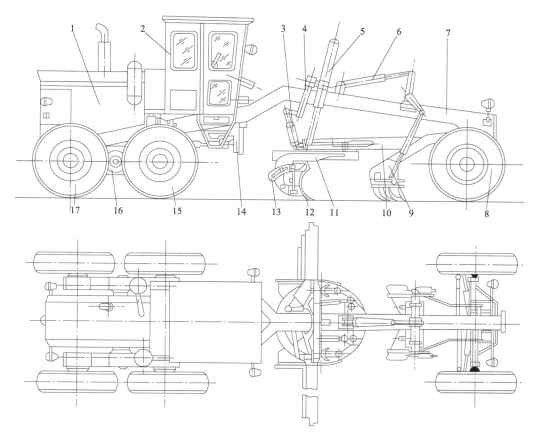

Fig. 2-5-1　F155 self-powered graders

1-engine;2-driver's cab;3-circle frame hydraulic cylinder;4-swinging mechanism;5-lifter cylinder;6-ripper stretch cylinder; 7-frame;8-front wheel;9-ripper;10-circle frame;11-blade circle;12-blade;13-angle position;14-transmission system;15-central wheel;16-tandem;17-rear wheel

　　The front and rear frames are rigid, welded steel construction. On the rear frame are mounted engine, gear box, tandems, hydraulic and fuel tanks, brake system, driver's cab. On the front frame are mounted front dozer, blade circle frame, saddle and hydraulic cylinders. Front and rear frames are connected by the articulation unit.

　　The central grader blade is the main working equipment of a grader. Because of a large range of cutting angles and mould board settings the cutting edge can be set to give the best scrape and cut qualities.

　　The grader blade is attached to the blade circle, this circle is turned by a hydraulic motor via a worm and wheel drive unit. The circle is attached to the circle frame (drawbar). [①] The circle frame is mounted to the front of the grader frame by means of swivel ball joint. Mounted on the back of the grader blade is a scarifier (ripper) with 6 scarifier teeth. It is possible to scarify in forward or reverse direction. The dozer blade is operated, i. e. raised or lowered by means of a hydraulic cylinder which is controlled by a hand operated valve in the main hydraulic valve block.

  The dozer attachment is designed for rough grader working. The back ripper is used for heavy scarifying work. It is equipped with scarifier teeth and operated by a hydraulic cylinder, this cylinder is controlled by a hand- operated valve in the main hydraulic valve block. The board ripper beam allows for ripper up to grader track width.

New Words

1. grader ['greidə]　　　　　　　　　　　n.　　　平地机
2. gradient ['greidiənt]　　　　　　　　　n.　　　斜度,斜率
　　　　　　　　　　　　　　　　　　　　adj.　　倾斜的
3. elevation [,eli'veiʃ(ə)n]　　　　　　　n.　　　提升,上升,高度,海拔
4. platform ['plætfɔ:m]　　　　　　　　n.　　　台,平台,站台,讲台
5. profile ['prɔfail]　　　　　　　　　　n.　　　侧面,外形,轮廓
6. manoeuverable [mə'nu:vərəbl]　　　adj.　　机动的,可调动的
7. sideways ['saidweiz]　　　　　　　　adv.　　斜着,斜向一边的
8. articulation [ɑ:,tikju'leiʃ(ə)n]　　　　n.　　　连接,接合,关节
9. drawbar ['drɔ:bɑ:]　　　　　　　　　n.　　　拉杆,牵引杆
10. swivel ['swivl]　　　　　　　　　　　n.　　　旋转
　　　　　　　　　　　　　　　　　　　　vt.　　　使旋转
11. logarithms ['lɔgəriθmz]　　　　　　　n.　　　对数
12. scarifier ['skeərifaiə]　　　　　　　　n.　　　松土机,翻路机
13. saddle ['sæd(ə)l]　　　　　　　　　　n.　　　支座

Phrases and Expressions

1. self-powered grader　　　　　　　　　　　　　自行式平地机
　　powerful self-propelled grader
2. prop shaft = propeller shaft　　　　　　　　　主传动轴
3. transfer box　　　　　　　　　　　　　　　　分动箱;分动器
4. swivel ball joint　　　　　　　　　　　　　　 转环球接头,滚珠回转接头
5. mould board　　　　　　　　　　　　　　　　刮板
6. steerable wheel　　　　　　　　　　　　　　　转向轮
7. driving wheel　　　　　　　　　　　　　　　　驱动轮
8. blade circle　　　　　　　　　　　　　　　　　环形刀架
9. circle frame (drawbar)　　　　　　　　　　　　牵引架
10. tandem chain　　　　　　　　　　　　　　　 平衡串联传动箱

Notes

①The circle frame is mounted to the front of the grader frame by means of a swivel ball joint.

Mounted on the back of the grader blade is a scarifier(ripper) with 6 scarifier teeth. It is possible to scarify in forward or reverse direction.

牵引架通过滚珠回转接头安装在平地机机架的前部。六齿松土器装在平地机刮刀的后面,可以向前或向后翻土。

Exercise one

Translate the following specific terms:

1. self-powered grader
2. direct injection engine
3. articulated frame
4. hydraulic cylinder
5. scarifier
6. the best scrape and cut qualities

Exercise two

Translate the following sentences into Chinese:

1. Self-powered graders are used in road construction to excavate side ditches, shape the surfaces and sides of fills and cuts and give them the required gradients.

2. The front and rear frames are rigid, welded steel constructions.

3. The grader blade is attached to the blade circle, this circle is turned by a hydraulic motor.

4. A self-powered grader is a highly manoeuverable machine, and its blade can be set at various angles both horizontally and vertically or brought out sideways.

Lesson 6

Asphalt Pavers

Asphalt pavers or bituminous finishers are special machines for paving asphalt mixture which are mixed by asphalt plants. Pavers can pave mixed asphalt mixture rapidly and uniformly on the road base, and can compact and level it.

The production process of an asphalt paver is as follows: the rear wheels of reversing dumper contact the push roller 12 of the asphalt paver and dump the asphalt mixture to the hopper 13 of the paver. Asphalt mixtures are delivered to the auger conveyor 8 by the drag conveyor 10 which has both left and right individually driven. Asphalt materials are further spread to two sides of the road. Tampers 7 compact them primarily. Screed unit 6 on which there are vibrator, heating unit, crowning device and screed plate makes the asphalt mixture form positive shape and thickness of mat. The components of asphalt paver are shown in Fig. 2-6-1.

1. Pushing Roller

When the material-carrying dumper in its idle position dumps the material into hopper of the paver, the pushing roller pushes it to move forward synchronistically with the paver until the material has been dumped up and the dumper goes away.

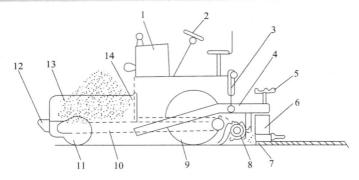

Fig. 2-6-1 Asphalt paver

1-engine;2-steering wheel;3-lifting hydraulic cylinder;4-side arm;5-thickness control;6-screed unit;7-tamper;8-auger conveyor;9-driving wheel;10-drag conveyor;11-front wheel;12-push roller;13-hopper;14-gate

2. Screed

The screed is the most important part of a paver;[①] it completes the paving sequence and produces the mat in its final state, to the required depth and with the necessary finish and a high degree of initial compaction. The screed may be equipped with either tampers or vibrators. The function of the tamper is to tuck the material under the leading edge of the screed, and at the same time to provide a certain amount of compaction in the material.

Where the machine is required to work mainly with crushed stone and dry lean concrete, the screed equipped with vibrators will handle these material very efficiently and produce an even texture finish.

The screed comprises two main sections to form the basic screed and extensions, which are built out to the required width. Each section or extension is robustly constructed and machined to ensure an accurate match and to produce a rigid, nonflexing unit.

The screed planning length provides each course with a smooth riding surface. To aid the flow of material under the vibrating screed, the screed shield is angled and an adjustable strike-off plate is incorporated. The screed may be adjusted to impart a crown of 3% to the mat. The screed equipment also includes glow plug ignition and an oil-fired heating system, to bring the screed to its working temperature and maintain it at that temperature. Some pavers set electricheating unit or propane burners.

3. The Material Hopper

The hopper, located at the front end of the paver, is designed to receive the mixed material dumped by the dump truck, and will hold sufficient materials to enable the paver to work during the changeover of delivery trucks. This ensures a continuity of action that is essential to obtain a smooth mat. At the bottom of the hopper, there are two bar feeders, and on the left and right hand sides, there are two movable walls which can be lifted or lowered by the action of the cylinder to enable all the materials in the hopper to be fed onto the bar feeder.

4. The Gate

On the front plates of the machine frame are mounted the left and right hand gates, which are designed to regulate the quantity of the mixed material fed from the hopper into the machine body. They are controlled by two cylinders respectively.

5. The Bar Feeder

In the machine, a two-row bar conveyor is applied. The two conveyors are controlled by air valves respectively. As the operating lever is pushed backward by the air valve, the feeding mechanism feeds material and as the lever is pushed forward, the mechanism ceases th feed material.

6. The Auger

The function of the auger is to spread the mixed material over the front edge of the smoothing screed. The structure of the auger is shown in Fig. 2-6-2.

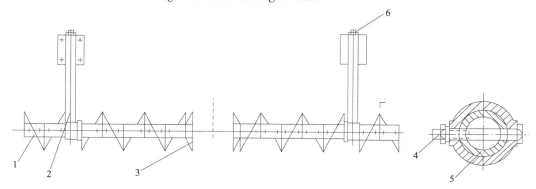

Fig. 2-6-2 The Auger
1-pitch of auger;2-support arm;3-flight;4-connection bolt;5-connection sleeve;6-back plate

The auger, which is connected to the machine frame by supports, has its left and right hand sections each of which is controlled by[②] the same air valve as used for the corresponding bar feeder. The flights of the auger are mounted in section with a pitch divided into two parts to be put together by use of connection sleeves and bolts.

Left-hand of auger is left rotation, right-hand of auger is right rotation.

At the ends of the right and left augers are mounted reversing flights to feed materials to the middle.

To assist the crewman to achieve accuracy of mat level, the paver has a complete system of automatic level control, which in any of the available combinations depends on controlling the tow points of the two side arms by electrical signals. The two side arms have to be controlled by independent means, either by a sensor on a datum, or a crossfall. A sensor can be adapted to accept a datum from a pre-set tensioned wire, a solid reference provided by the kerb, haunch or adjacent mat, or an averaging beam which averages the level of a road over a distance of 6 m to 12 m. Automatic level control is achieved by combining any two of these methods to control the side arm tow points.

An asphalt paver is usually powered by a diesel engine resiliently mounted to prevent vibration being transmitted to other parts of the paver. The four pumps which provide the independent hydraulic supplies for traction, the two conveying systems, tamping and/or vibrator drives and ram services are mounted on a splitter gearbox direct coupled to the engine.

Asphalt pavers running gear can be crawler-mounted or wheeled-mounted.

New Words

1.	bituminous [bi'tjuːminəs]	adj.	(含)沥青的
2.	asphalt ['æsfælt]	n.	沥青
3.	crowning ['krauniŋ]	n.	凸面;拱起
4.	mat [mæt]	n.	垫子,铺层
		vt.	给……铺上
5.	screed [skriːd]	n.	整平板
6.	sequence ['siːkwəns]	n.	顺序,程序
7.	tuck [tʌk]	vt.	卷起,塞
8.	concrete ['kɔnkriːt]	n.	混凝土
9.	texture ['tekstʃə]	n.	结构,组织
		adj.	有形的,具体的
10.	non-flexing [ˌnɔn'fleksiŋ]	adj.	非弯曲的,非挠性的
11.	vibrator [vai'breitə]	n	振动器
12.	tamper ['tæmpə]	n.	夯,振捣棒(器)
13.	shield [ʃiːld]	n.	护板,盾
14.	glow [gləu]	n. ;vi.	辉光,发辉
15.	propane ['prəupein]	n.	丙烷
16.	synchronistically [ˌsiŋkrəu'nistikəli]	adv.	同步
17.	regulate ['regjuleit]	vt.	调解,控制
18.	two-row [tuː'rəu]	adj.	两排的
19.	conveyor [kən'veiə] (conveyer)	n.	输送机,输送带
20.	sag [sæg]	n. ;vi.	下垂
21.	auger ['ɔːgə]	n.	螺旋输送机,螺旋布料器
22.	flight [flait]	n.	螺旋片,飞行
23.	pitch [pitʃ]	n.	斜度,螺距,齿节
24.	sleeve [sliv]	n.	套管,套筒
		vt.	给……装套筒(管)
25.	alter [ɔːltə]	vt. ;vi.	改变,变动

Phrases and Expressions

1. asphalt paver   沥青摊铺机
2. bituminous finisher   沥青路面加工机
3. even texture finish   均匀的路面

Part 2 Engineering and Construction Machineries

4. glow plug ignition　　　　　　　　火花塞点火
5. strike-off plate　　　　　　　　　整平板
6. propane burner　　　　　　　　　丙烷燃烧器
7. bar feeder　　　　　　　　　　　刮板输送机,带式进料器

Notes

① ...it completes the paving sequence and produces the mat in its final state, to the required depth and with the necessary finish and a high degree of initial compaction.

……它完成摊铺工序并形成带料的最终状态,达到必要的整平度和初压实高度。

② ...the same air valve as used for the corresponding bar feeder.

……与用来控制相应的刮板输送器是同一个气阀。

as used for...可看作 as it is used for...,作定语从句,修饰 air valve。

Exercise One

Put the following expressions into Chinese:

1. the quality of the laid mat
2. screed and extension
3. asphalt paver
4. mixed material
5. material hopper
6. lifting hydraulic cylinder

Exercise Two

Translate the following sentences into Chinese:

1. On receiving the material from the conveyors, the chain driven auger spreads the material evenly in the front of the leading edge and across the complete width of the screed.
2. The material is conveyed from the bottom of the hopper via a pair of flow gates to the auger box.
3. The operator can easily observe the whole auger channel and hopper as well as the feeding truck and the direction indicator.
4. An asphalt paver is usually powered by a diesel engine resiliently mounted to prevent vibration being transmitted to other parts of the paver.

Lesson 7

Cement Concrete Pavers

Concrete has widely been applied in road building. It has also widely applied machinery to make concrete road surface. These machines may be classified into two groups: rail type or paver train and slipform paver.

Modern concrete pavers are mostly slipform pavers because of high productivity and compact structure. With the slipform paver working from a near perfect grade, the yield is high and the surface deviations can be held to a minimum. In recent years, the technique of slipform paving has developed to the point where it is providing better roads at low costs per square meter than any other method.

The slipform paver has the paver, weight and positive traction required to handle big loads and lay slabs with a square, vertical edge, in a perfectly straight line at high production rates. It is almost impossible for the slipform paver to spin out or get stuck because of the unique track flow divider system.

The slipform paver can either form a slab over steel or without steel. It can distribute and meter material dumped on grade ahead of it, or condition and finish a slab initially formed with a placer spreader.

The SF-350 four track slipform paver is shown in Fig. 2-7-1.

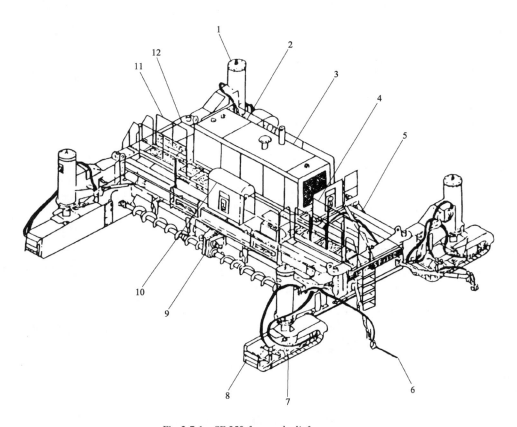

Fig. 2-7-1　SF-350 four track slipform paver

1-leg assemblies; 2-oil tank; 3-power assembly; 4-passage; 5-telescoping frame; 6-automatic leveling pick-up; 7-steering pick-up; 8-track assembly; 9-paving unit; 10-console assembly; 11-hand rail; 12-boundary beam

During operation, concrete that is dumped in front of the machine is spread evenly across the front of paving unit by augers. An adjustable strike-off behind the augers then allows only a preset amount of concrete to pass under the strike-off to the vibrators and tamper bars just ahead of the profile pan. [1]The vibrators fluidize the concrete for good compaction and easier forming, while the tamper bar pushes the large aggregate just below the surface so the profile pan can form a uniform top surface. The float pan follows the profile pan to provide the finished slab top surface.

Side forms at each side of the profile pan form the sides of the slab and edges at each side of the float pan form the finished edges of the slab.

Machine functions are performed by five separate hydraulic systems: the ground drive system, auger drive system, vibrator system, tamper bar drive system and auxiliary system.

Power for all systems is provided by the primary power system, and machine propulsion is accomplished by four track assemblies.

The single lane version of the slipform paver offers the same advantages as the dual lane machine, but is able to work in more confined areas. It is ideal for widening existing roads, work on ramps, medians, interchanges or other short radius operation.

②The dual-lane autograde slipform paver employs a 6-step methods of conditioning, forming and finishing a slab that produces unrivaled quality and accuracy.

6-step continuous paving:

(1) spreader-auger distributes mix.

(2) primary feed meter regulates flow of mix to the vibrator compartment.

(3) internal vibrators consolidate material.

(4) secondary feed meter reapportions and vibrates mix while circulating grout around aggregate before mix reaches screed.

(5) dual-oscillating extrusion screeds (primary and secondary) consolidate the mix, form the initial slab profile and make the initial finish.

(6) floating finisher with the edging unit shapes the slab to the final conformity.

The production process of the slipform paver includes auger, strike-off plate, vibrators, tampers, profile pan, bar inserting device, crowning, float pan, edges, mopping and so on (Fig. 2-7-2).

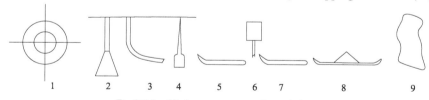

Fig. 2-7-2 Slipform paver processing technique

1-auger; 2-strike-off; 3-vibrator; 4-tamper; 5-profile pan; 6-bar inserting device; 7-float pan; 8-edge; 9-mopping

New Words

1. concrete [ˈkɒŋkriːt]　　　　　　n.　　　　　混凝土
2. slipform [slipˈfɒm]　　　　　　n.　　　　　滑模造型
3. productivity [prɒdʌkˈtɪvɪtɪ]　　n.　　　　　生产率, 生产量
4. yield [jiːld]　　　　　　　　　vt.　　　　　产生, 给予
　　　　　　　　　　　　　　　vi.　　　　　出产, 服从
　　　　　　　　　　　　　　　n.　　　　　产品, 产量

5. deviation [diːviˈeiʃ(ə)n]   n.   偏差,背离
6. slab [slæb]   n.   平板,厚板,大理石
7. fluidize [ˈflu(ː)idaiz]   vt.   使流体化
8. aggregate [ˈægrigət]   n.   集料,骨料
9. profile [ˈprəufail]   n.   侧面,外形
10. lane [lein]   n.   车道
11. ramp [ræmp]   n.   斜道,斜坡
12. median [ˈmiːdjən]   n.   中间道,中间分隔带
           adj.   中间的,中线的,中灰色的
13. interchange [intəˈtʃein(d)ʒ]   vt.; vi.   交换,交替
           n.   交叉路口
14. consolidate [kənˈsɔlideit]   vt.   加固
15. reapportion [riːəˈpɔːʃ(ə)n]   vt.   重新分配
16. grout [graut]   n.   石灰浆,水泥浆
17. extrusion [eksˈtruːʃən]   n.   挤压,挤压成型
18. conformity [kənˈfɔːmiti]   n.   依从,遵照,要求
19. edge [edʒ]   n.   磨边器,边磨板
20. mop [mɔp]   n.   拖布,抹布
           vt.   用布擦,用布拖

Phrases and Expressions

1. slipform paver   滑模式摊铺机
2. cement concrete paver   水泥混凝土摊铺机
3. rail type or paver train   轨道式摊铺机
4. spin out   拖延,消磨
5. get stuck   被……堵
6. profile pan   成型盘
7. float pan   浮动盘
8. side form   侧模板
9. strike-off   虚方控制板,整平板
10. bar inserting device   拉杆插入器

Notes

①The vibrators fluidize the concrete for good compaction and easier forming, while the tamper bar pushes the large aggregate just below the surface so the profile pan can form a uniform top surface.

振动器提浆以使混凝土密实和容易成型;而振捣棒则将大集料挤到面层下方以便成型盘形成均匀表面层。

Part 2 Engineering and Construction Machineries

②The daul-lane autograde slipform paver employs a 6-step methods of conditioning, forming and finishing a slab that produces unrivaled quality and accuracy.

双车道自动找平滑模式摊铺机采用了六步法调整,成形和整平混凝土,从而获得无与伦比的质量和精度。

Exercise One

Put the following expressions into Chinese:

1. slipform paver
2. cement concrete paver
3. rail type or paver train
4. profile pan
5. float pan
6. side form

Exercise Two

Translate the following sentences into Chinese:

1. The vibrators fluidize the concrete for good compaction and easier forming, while the tamper bar pushes the large aggregate just below the surface so the profile pan can form a uniform top surface.

2. The daul-lane autograde slipform paver employs a 6-step methods of conditioning, forming and finishing a slab that produces unrivaled quality and accuracy.

3. During operation, concrete that is dumped in front of the machine is spread evenly across the front of paving unit by augers.

4. Modern concrete pavers are mostly slipform pavers because of high productivity and compact structure.

Lesson 8

Rollers

①Loose soil placed in an earth structure must be compacted as layer after layer is added to attain the required density and stability. Soil may be allowed to compact by itself, but present day rates of road construction almost entirely exclude this natural compaction and the work is generally done by machine. Compactors are also used to compact undisturbed soil, as it occurs in nature, for example at the bottom and on the slopes of cuts and channels in order to reduce water lose by filtration.

Compactors are also employed when building roads by the soil stabilization method.

Rollers compact the soil by the pressure exerted by their rolling wheels on the surface. They are the most widespread and simplest machines for soil compacting. They may be of static smooth, sheep-foot, air-tyred(pneumatic type) or vibration design.

The static smooth rollers which compact the soils by its weight can compact road-beds、road surfaces、squares and road foundations of a variety of the constructions.

It can be classified as : two-wheel and two-axle type; three-wheel and two-axle type; three-wheel and three-axle type.

The pneumatic type rollers are designed to meet the requirements for rolling and compacting macadam and bitumen roads as well as asphalt concrete road foundations in capital constructions, civil engineering work and national defence work. Through the adjustment of ballast and the variation of inflation of the tyres better rolling performance can be obtained in the compaction of sandy or sticky soil. During the compaction the grain sizes of the paving material will not be altered and a better cohesion between different layers of pavement can be secured.

The 4-front-wheels (steering wheels) as shown in Fig. 2-8-1 can be moved up and down, the 5-rear-wheels (driving wheels) as shown in Fig. 2-8-2 are driven by the two sprockets each at the left and right wheel. The front and rear wheels dispose staggeringly in order to compact the parts not being rolled.

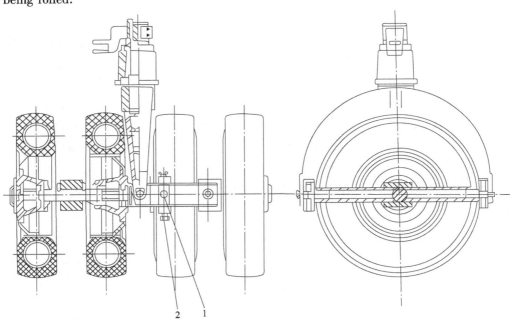

Fig. 2-8-1 Front wheels
1-pin;2-bolt

The water sprinkling system is mounted on this machine, and sprinkles according to the requirement of the road surface.

Counter-balance iron piece are also provided beneath the body for adjustment of the weight of the road roller.

The single drum vibratory rollers of series BW212 are self-propelled soil compactors for use in earthwork. Compacting by vibration is achieved by the reciprocating movements of a vibrator, which shift the soil particles and compact the soil. Versions differ primarily in compaction capacity and climbing ability.

Part 2 Engineering and Construction Machineries

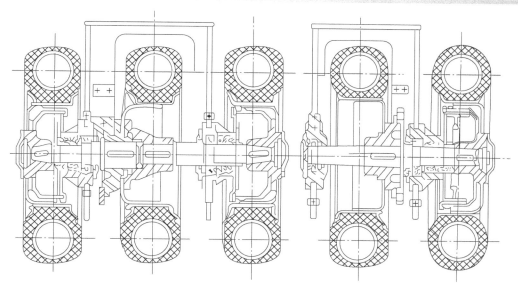

Fig. 2-8-2   Rear wheels

All of them are equipped with a diesel engine which drives the hydraulic pumps for travel drive, vibration and steering at constant speed.

The frame of the vibratory roller consists of front frame with scrapers and rear frame, both frames being connected by an articulated joint. The articulated joint can be blocked by a travel lock.

The drum with vibration motor on the left-hand side and with scrapers is flexible suspended in the front frame.

The rear frame supports the diesel engine with hydraulic pumps for travel drive, vibration drive and steering, the hydraulic oil tank, the fuel tank, driver's cabin, driver's panel, driver's seat on the battery box, and the travel motor for the drive wheels. The rear frame rests on the drive axle.

The diesel engine drives the travel pump (variable pump) which provides the hydraulic drive for the travel motor of the rear drive wheels. The drive wheels are rubber tyred and filled with water and calcium (or magnesium) chloride to increase the weight; air pressure applied additionally.

The travel pump for the travel wheels also drives the drive motor for the drum drive.

The diesel engine drives the vibration pump via a flexible coupling. The vibration pump provides the hydraulic drive for the vibration motor on the left-hand side of the drum, which generates the vibration by rotating the eccentric shaft.

Vibration is turned on by means of a switch on the left side of the driver's panel.

Vibration is allowed at full engine speed only.

The vibration pump is variable displacement pump with two flow directions. Change of vibration frequency and amplitude is achieved by reversing the pump flow which changes the sense of rotation of the vibration motor. Rotation of the vibration motor can change the relative positions of the fixed eccentric weight and movable eccentric weight, so changing the eccentric mass (overlay or offset) and amplitude. High frequency is connected with small amplitude, low frequency with a large amplitude.

The frequency selector is used to preselect either high or low frequency, its center position is neutral (i.e. no frequency selected).

Power steering is supplied with pressure oil from steering pump (driven by the diesel engine).

Braking during travel is effected via travel drive control (hydrostatic brake). The parking brake is a manually operated disc brake which acts as on the drive shaft of the drive axle.

Layout of vibratory roller is shown in Fig. 2-8-3.

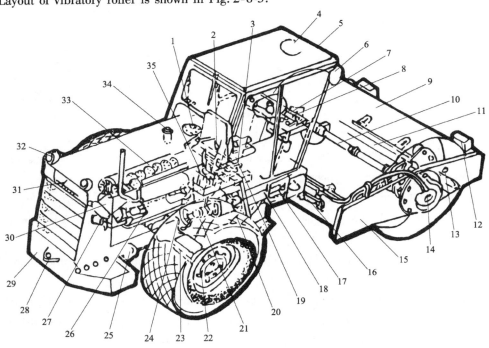

Fig. 2-8-3 Layout of BW 212 vibratory roller

1-hydraulic oil filler neck;2-sight glass for hydraulic oil tank;3-pull knob for heating;4-ventilator;5-operator's cabin;6-rear mirror; 7-vibration motor;8-travel control lever;9-drum;10-eccentric shaft;11-scraper;12-heat light;13-lifting eye of front frame;14-drum drive motor;15-front frame;16-articulated joint lock;17-articulated joint;18-steering cylinder;19-throttle lever;20-travel motor for drive axle;21-drive shaft;22-planetary;23-drive axle;24-parking brake;25-lifting eye of rear frame;26-steering pump;27-vibrating pump; 28-tow eye;29-rear frame;30-dry air filter;31-tool compartment;32-working light;33-diesel engine;34-fuel tank filler neck;35-wedge

New Words

1. compactor[kəmˈpæktə]　　　　　n.　　　　　压实机械,夯具
2. exclude[iksˈkluːd]　　　　　　　vt.　　　　　排除,拒绝接纳
3. filtration[filˈtreiʃən]　　　　　　n.　　　　　过滤,过滤作用
4. macadam[məˈkædəm]　　　　　n.　　　　　马克当(碎石)路面
5. vibrator[vaiˈbreitə]　　　　　　n.　　　　　振动器
6. earthwork[ˈəːθwəːk]　　　　　　n.　　　　　土方工程

Part 2 Engineering and Construction Machineries

7. block[blɔk]	n.	大块,阻塞
	vt.	阻塞,闭锁
8. suspend[səs'pend]	vt.	吊,挂
	adj.	悬挂的;吊式的
9. eccentric[ik'sentrik]	adj.	偏心的,离心的
	n.	偏心轮
10. planetary['plænit(ə)ri]	adj.	行星的;行星齿轮的
11. variable['veəriəb(ə)l]	adj.	可变的,可调的
12. frequency['fri:kwənsi]	n.	频率,次数
13. amplitude['æmplitju:d]	n.	振幅,幅度
14. calcium['kælsiəm]	n.	钙
15. magnesium[mæg'ni:ziəm]	n.	镁
16. chloride['klɔ:raid]	n.	氯化物
17. mercury['mə:kjuri]	n.	水银

Phrases and Expressions

1. layer after layer —— 一层一层地
2. static smooth roller —— 静力光轮压路机
3. sheep-foot —— 羊角碾
4. air-tyred(pneumatic type) —— 轮胎压路机
5. vibration roller —— 振动压路机
6. planetary drive —— 行星传动
7. eccentric shaft —— 偏心轴
8. bitumen roads —— 沥青路面
9. asphalt concrete —— 沥青混凝土
10. road foundation —— 地基
11. grain sizes —— 粒度
12. fixed eccentric weight —— 固定偏心块
13. movable eccentric weight —— 移动偏心块

Notes

①Loose soil placed in an earth structure must be compacted as layer after layer is added to attain the required density and stability.

土工建筑中的松土必须一层一层地压实,层层叠加达到所要求的密实度和稳定性。

本句中 as 引起状语从句;to attain 是主句的目的状语。

Exercise One

Put the following expressions into Chinese:

1. the high dynamic forces applied to the material

2. the vibration produced by a rotating eccentric weight inside the drum

3. the drive wheels filled with water

4. fixed eccentric weight

5. movable eccentric weight

6. static smooth roller

Exercise Two

Translate the following sentences into Chinese:

1. All of rollers are equipped with a diesel engine which drives the hydraulic pumps for travel drive, vibration and steering at constant speed.

2. The term frequency refers to the number of vibrations of the drum per second.

3. By amplitude we mean the amount by which the vibrating drum moves "up" and "down" during the compaction process.

4. Compactors are also employed when building roads by the soil stabilization method.

Lesson 9

Asphalt Plants

Asphalt mixture is widely used in the road making. It is a high grade material of road surface. At present producing of asphalt mixture has been mechanized. Its productive equipment is called "asphalt mixture plant".

Asphalt mixture plant can be classified as follows:

(1) By the period of operation and the area serviced, asphalt mixing plants can be subdivided into stationary, semi-stationary and mobile.

(2) By their productive process: conventional (or batch) and continuous (or drum).

1. Basic Batch Mix Layout

As shown in Fig. 2-9-1, from the sand hopper by belt feeder and from the stone hopper by either electric vibrating feeders or belt feeders, material is fed onto a continuous belt conveyor, and delivered to the dust plant/dryer via an inclined belt conveyor.

The belt conveyor delivers material to the dryer feed ring, which moves the material onto a series lifters.[①] As the dryer cylinder rotates, the material is cascaded through hot gases produced by the oil burner and travels along the length of the cylinder to the discharge end. Pusher plates allow side discharge of material into a discharge chute, which feeds into the boot of the hot stone elevator. The multi-cyclone, single stage dust collecting plant mounted on the rear of the chassis collects the dust and is delivered to the dryer discharge chute by means of two screw conveyors. The hot stone elevator takes the material from the dust plant/dryer unit onto the vibrating screen where the material is graded and delivered into individual compartments of the hot stone storage bins. Oversize material rejected by the top deck of the screen is passed to a reject chute at the end of mixing and screen section.

Part 2 Engineering and Construction Machineries

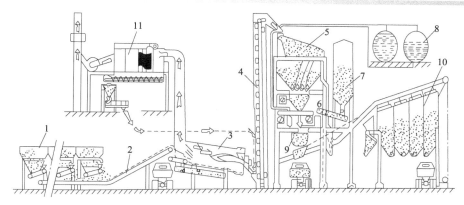

Fig. 2-9-1 Batch mix layout

1-cold feed unit;2-belt conveyor;3-dryer cylinder;4-hot stone elevator;5-screen unit;6-hot stone metering device;7-filler silo;8-bitumen system;9-mixer;10-storage bin;11-dust plant

The filler is loaded into the silo and with the binder supplied in circulation, the plant is ready for batch production.

From a control cabin the operator controls the flow of the material according to the specification required. Aggregate is discharged to the batch weigh hopper from the storage bins and weighed. The material from the batch weigh hopper is then discharged into a fully enclosed batch elevator and conveyed to the mixer. Filler is added from the filler silo through a filler weight hopper adjacent to the mixer. Binder is weighed into a binder weigh tank adjacent to the mixer. Aggregate is passed into the mixer for preset dry mixing time after which binder is added and mixing continues for the necessary wet mix period until the batch is ready for discharge into the transporter standing below the discharge doors of the mixer. Provision is made for filler to be added before or after binder according to requirements.

2. Cold Feed Unit

②The cold feed unit is designed for loader type feeding and consists of individual sand and stone hoppers each complete with its supporting structure and feeder bolted together to form one unit.

Sand hoppers are fitted with belt feeders and stone hoppers can be fitted with either belt feeders or vibrating tray feeders. The belt feeders can also incorporate variable speed drives. These units deliver material onto a horizontal trough collecting conveyor transporting material to a link conveyor which joins the cold feed unit to dryer.

Adjustable doors are fitted to outlet of each hopper to regulate the feeder rate and to facilitate the removal of any obstruction.

Feeders and conveyors are each fitted with individual electric motors for driving.

3. Dryer/Dust Plant

The unit incorporates an inclined link conveyor for delivering material from the cold feed unit to the dryer drum by means of a wide trough belt. The dust plant incorporates multi-cyclone single stage dust plant and utilizes a paddle blade type fan driven by an electric motor. The collected dust is delivered to the discharge chute by means of two screw conveyors built into the unit or alternatively it can be blown to a silo for a controlled addition to the mixer.

③The dryer cylinder is constructed of heavy rolled steel plate and fitted with flights of lifters staggered along the length of the cylinder. The cylinder is fitted with two machined steel support bands mounted on brackets, with a preset allowance for expansion. A feed drum receives the cold aggregate from the link conveyor and elevates it into the cylinder into a series of lifters which cascade the material through the hot gases produced by the oil burner.

As the material reaches the end of the cylinder, pusher plates allow side discharge of material into discharge boot which feeds into the boot of a hot stone elevator.

The drum is driven by chain drive through an electric motor and totally enclosed gear unit, a chain ring is fixed around the drum and a chain pinion incorporated in the electric motor and gear unit assembly. The drive assembly incorporates a soft start mechanism to give a very smooth start, which results in a considerable increase in the life of the chain, chain wheel and chain pinion.

Heat is generated by a self-proportioning oil burner, situated at the delivery end of the cylinder. ④The burner is a self-proportioning air/oil ratio and is complete with combustion chamber, blowing fan, oil pump, piping, filters, valve and gauges in a package unit designed to the very latest specification and incorporating a retractable carriage to enable the whole unit to be withdrawn from the drum for ease of maintenance. Remote automatic ignition and manual flame modulation with temperature indication, together with fan and flame failure indicators are supplied as standard equipment. The dryer discharge chute incorporates an infra-red type pyrometer with remote temperature indication mounted in the control cabin.

4. Mixing/Screening Unit

The mixing and screening unit incorporates a hot stone elevator which receives material from the dryer through the dryer discharge chute into the elevator feed boot.

Buckets collect the material at the feed boot and elevate it to the top of the mixing and screening unit.

The hot storage bin section provides 3-5 compartments, each incorporating an overflow chute for delivering excess material to ground level.

Each bin compartment is fitted with tray type doors on self lubricating rollers and operated by two pneumatic cylinders, which enable the doors to be closed in two stages giving added accuracy of weighing. The batch weigh hopper is suspended on electrical load cell and delivers material to a fully enclosed batch elevator through a full width quadrant door operated through a pneumatic cylinder.

The batch elevator incorporates two runs of chain for conveying material from batch weigh hopper to the mixer.

Mixer is lined throughout with wear resistant line plates. It has cast steel paddle arms with renewable alloy tips. The paddle arms are secured to twin paddle shafts running in spherical roller bearings. The mixer is driven through two heavy duty chain wheels from a gear unit and electric motor. The whole drive unit incorporates an overload device to give protection against double batching.

The mixer incorporates a full width rotary door operated by pneumatic air cylinders, which ensure clean and efficient discharge of material from the mixer. The mixer has a mild steel dust cover to prevent fume and dust emission during mixing.

5. Filler System

The filler weigh hopper is situated above the mixer and suspended on electrical load cells. The filler weigh hopper discharges material through a pneumatically operated quadrant door.

The filler silo of capacity should be matched to plant output and specification requirements. The filler is fed into the filler weigh hopper from the filler silo through a screw conveyor. The screw conveyor is driven through a geared electric motor and flexible coupling. One or two independent silos can be supplied depending on customer requirements.

6. Bitumen System

Binder weigh tank is situated above the mixer and suspended on electrical load cells.

The bitumen tanks can be electrically or hot oil heated and are thermostatically controlled with capacities according to plant output and binder grade requirements. Bitumen is delivered to a binder weigh tank from the bitumen tank. The binder weigh tank is thermostatically heated. Binder is delivered from the weigh tank through a pneumatically operated valve into the bitumen feed pipe, which is connected to the bitumen spray bar supported on the mixer. The spray bar discharges binder to the mixer.

7. The Mixer Conversion Package

The mixer conversion package makes your batch or drum mixer plant a unique hybrid, incorporating the best features of both batch and drum types of plants.

Using the mixer conversion package provides the high-quality hot mix you expect of a batch plant coupled with the high yields of drum mixer production.

You get the very best of both facilities, while completely avoiding the problems inherent in each type of plant. It turns a batch plant into a continuous production facility, just like a drum mixer. When installed in a drum mixer plant, it separates the mixing and the drying processes, enabling you to generate premium quality hot mix at maximum speed.

New Words

1. mobile ['məubail]	adj.	移动的;运动的,装在车上的
2. conventional [kən'venʃ(ə)n(ə)l]	adj.	常规的;传统的,标称的
3. batch [bætʃ]	n.	一次生产量
	adj.	间歇(式)的
4. temporary ['temp(ə)rəri]	adj.	暂时的;临时的
5. dust [dʌst]	n.	灰尘,粉末
6. incline [in'klain]	vt. ;vi.	(使)倾斜,(使)倾向
7. cascade [kæs'keid]	n.	小瀑布
	vi.	瀑布式流动
8. chute [ʃuːt]	n.	滑槽,斜槽

9. boot[buːt]	n.	进料斗
10. multi-cyclone[mʌltiˈsaikləun]	n.	多管式旋风除尘器
11. reject[riˈdʒekt]	vt.	去掉,弃,拒绝
12. binder[baind]	n.	黏合剂,结合剂
13. adjacent[əˈdʒeis(ə)nt]	adj.	邻的;相近的
14. silo[ˈsailəu]	n.	料仓
15. aggregate[ˈægrigət]	n.	集料
	adj.	聚集的
	vt.;vi.	(使)聚集
16. self-proportioning[self-prəˈpɔːʃniŋ]	n.	自动配比
17. stagger[ˈstægə]	vt.	摆动;参差,交错
18. withdraw[wiðˈdrɔː]	vt.	退回;缩回,取消,移开
19. infra-red[ˈinfrə-red]	adj.	红外线的
20. pyrometer[paiˈrɔmitə]	n.	高温计
21. quadrant[ˈkwɔdrənt]	n.	扇形齿轮,四分之一圆
22. thermostatically[ˌθəːməsˈtætikli]	adv.	热控地
23. bitumen[ˈbitjumin]	n.	沥青
24. hybrid[ˈhaibrid]	n.	混血儿

Phrases and Expressions

1. asphalt plant	沥青混合料拌和设备
2. small scale site	小型工地
3. temporary occasion	临时工地
4. storage bin	贮料仓
5. top deck	顶层
6. reject chute	弃料槽
7. batch weigh hopper	称量斗
8. adjacent to	与……相邻
9. filler silo	粉料仓
10. vibrating tray feeder	振动给料槽
11. self-proportioning oil burner	自动配比燃油器
12. electrical load cell	电子载荷传感器
13. tray type door	蝶形门
14. quadrant door	蝶形门
15. paddle blade type fan	叶片式风扇
16. support band	滚圈
17. feed drum	(烘干筒)受料区
18. side discharge	(烘干筒)卸料端

Part 2　Engineering and Construction Machineries

19. paddle arm	拌和臂
20. mild steel	低碳钢

Notes

①As the dryer cylinder rotates, the material is cascaded through hot gases produced by the oil burner and travels along the length of the cylinder to the discharged end.

随着烘干筒的转动,集料在燃烧器产生的热气中散落下来,并沿着烘干筒长度方向移动到卸料口一端。

cascade 实为"像瀑布一样落下",这种现象在烘干工艺中称"料帘"。

②The cold feed unit is designed for loader type feeding and consists of individual sand and stone hoppers each complete with its supporting structure and feeder bolted together to form one u-nit.

冷集料供应装置是为装载式供料设计的,并由分离的砂料仓和石料仓组成,每个料仓都带有各自的用螺栓连在一起的支撑结构和出料机以形成一个整体。

③The dryer cylinder is constructed of heavy rolled steel plate and fitted with flights of lifters staggered along the length of the cylinder.

烘干筒由重型轧制钢板制成,并且沿着纵向交错安装有提升叶片。

④The burner is a self-proportioning air/oil ratio and is complete with combustion chamber, blowing fan, oil pump, piping, filters, valve and gauges in a package unit designed to the very latest specification and incorporating a retractable carriage to enable the whole unit to be withdrawn from the drum for ease of maintenance.

自动配比燃烧器与按最新规范设计的燃烧室、鼓风机、油泵、管道、滤清器、阀和仪表做成一体,整个装置包括一个可伸缩的托架以保证在维修时整个装置可以从烘干筒处移开。

Exercise One

Put the following expressions into Chinese:
1. asphalt plant
2. small scale site
3. temporary occasion
4. storage bin
5. self-proportioning oil burner
6. electrical load cell

Exercise Two

Translate the following sentences into Chinese:
1. Each mix specification is set up manually by the operator through the basic module control unit.
2. The filler silo of capacity should be matched to plant output and specification requirements.
3. In addition to foregoing batch asphalt plants in recent times the drum mixing plants have obtained widespread application.
4. The advantages of this drum mixing plant are simplifying productive technological process

and equipment, thus reducing cost and environmental pollution.

Lesson 10

Cold Milling Machines

Milling machines include hot milling machines and cold milling machines (Fig. 2-10-1). Hot milling machines with heating devices are usually used on operation of hot recycling road.

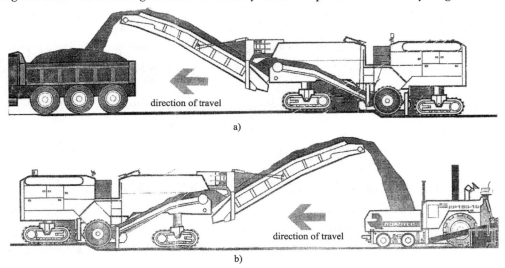

Fig. 2-10-1 Cold milling machines
a) Front loading up cutting; b) Rear loading down cutting

The cold milling machines are one of the asphalt concrete road maintenance equipment, mainly used in the renovation constructions of large-scale highways, urban roads, airports, freight yards etc..

The cold milling machines are used to remove the defective paving of upheavals, waves, oil layers, tracks and ruts to the required thickness so that it can be replaced and repaired. They are also used for trench and tunnel excavations, cement road surface roughening and surface stair level milling. The reclaimed asphalt pavement can be reused in stationary mixing plants without requiring further treatment.

The product range currently encompasses a variety of types of milling machinery and is designed to meet all the various requirements for processing pavement, from partial repair of areas to complete removal of entire road structures.

The cold milling machines consist of power unit, chassis, milling drum assembly, milling depth control (Automatic leveling system), hydraulic system, electric system, loading milled material system and other auxiliary equipment.

The milling machines can be classified as front loading up cutting and rear loading down cutting types.

Part 2 Engineering and Construction Machineries

1. Basic Design

The (W2100) milling machine is crawler-typed and mechanically driven milling drum and a two-stage front loading conveyor system, which can be adjusted in height and slewed to both sides.

2. Power Unit

The machine is driven by a modern diesel engine. The engine complies with the stringent requirements of the exhaust emission standards stipulated by US Environmental Protection Agency and European Union. It is equipped with a fully electronic engine management system which allows the engine to automatically adapt to varying ambient conditions, such as changing atmospheric pressure, temperature or humidity. In addition, the engine offers maximum torque stability even at extreme engine loads, thus preventing breaks in operation.

3. Milling Drum (Fig. 2-10-2)

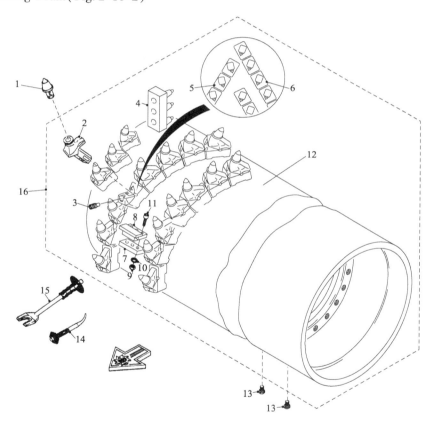

Fig. 2-10-2 Milling drum

1-milling cut; 2-milling cut upper holder; 3-retaining bolt; 4-combination cut tool; 5、6-displacement of milling cut tool; 7、8-milling cut buttom holder; 9-nut; 10-coil pad; 11-connecting bolt; 12-milling drum; 13-bolt; 14、15-mixing cut tool; 16-cover

The milling drum operates in up-milling direction. Toolholders accommodating the point-attack cutting tools are welded onto the drum body. Special edge segments ensure a clean sharp cut at the edges. Additional ejectors ensure that the milled material is efficiently transferred to the primary conveyor. If the milled material is to remain on the ground, a flap at the scraper blade ensures that it is deposited between the crawler tracks. As an option, the milling drum can be equipped with patented and established quick change toolholder system. With this system, the bottom parts of toolholders are welded onto the drum body. The upper parts are secured to the bottom parts by retaining bolts to allow quick replacement.

The milling drum is driven mechanically by the diesel engine via a mechanical clutch and power belts acting on the drum gearbox. The power belts ensure optimum power transmission due to their width and the use of 12 V-belt, and have a long service life. Constant tension of the power belts is automatically maintained by a hydraulic cylinder.

4. Cutting Tool Replacement

①The scraper blade opens hydraulically (can be slewed by 100°), thus providing good access to the milling drum for replacement of cutting tools. Sufficient depositing space for the cutting tool containers is provided at both rear crawler tracks.

5. Loading Milled Material

Loading of the milled material from the milling chamber on trucks is effected to the front (front loading) by means of a wide conveyor system consisting of a primary conveyor and a discharge conveyor.

A gradation control beam largely prevents the asphalt from breaking into slabs and simultaneously protects the primary conveyor against premature wear and tear. ②The discharge conveyor can load trucks from a great height, its height is adjustable and can be slewed to both sides, thus always allowing an optimum adaptation to the conditions prevailing on site.

The high conveying speed and the generously dimensioned, wide, V-ribbed steep-incline belts ensure that the material is quickly removed from the drum housing.

The discharge conveyor is covered to prevent clouds of dust from being blown away by the wind and causing a nuisance. The design of the conveyor system allows an easy replacement of the belts.

6. Milling Depth Control/Automatic Leveling System

The milling machine is equipped with an electronic automatic leveling system to control the milling depth. It is governed by means of proportional control, meaning that changes in the height of the reference plane are compensated quickly and without overshooting of the machine.

③The reference planes can be scanned by various methods, for instance, by a wire-rope sensor at the side plates, an ultrasonic sensor on the existing road surface, a grade-line in combination with rotary transducers, or a plane formed by lasers.

A slope control sensor is available as an equipment option; the required connections are included as a standard feature.

7. Travel System and Drive

The crawler track are suspended on the chassis by means of cylindrical with hydraulic height adjustment. The height of each crawler track can be adjusted individually. The milling depth is set via the two front columns, while the rear crawler tracks act as a full floating axle. The large stroke

Part 2　Engineering and Construction Machineries

provides a large ground clearance, thus facilitating even difficult manoeuvres, such as revering in the milled track or loading the machine onto a low-bed trailer.

The milling machine is equipped with large crawler tracks. Each crawler track is driven by its own hydraulic motor. The travel drive motors are fed by a common hydraulic variable displacement pump.

The crawler tracks are driven automatically, thus dispensing with the need to change between milling and travel gear. The speed can be infinitely varied from zero to the high maximum speed.

A switchable hydraulic flow divider acts as differential lock and ensures uniform traction even under difficult conditions.

8. Steering System

The machine has a finger-light hydraulic all-track steering system, which can be operated from both the right and the left side of the operator's platform. It is governed by means of proportional control, and the front and rear tracks are steered separately via joystick.

9. Brake System

Braking is achieved by the self-locking hydrostatic transmission. The road milling machine is additionally equipped with two automatic multiple-disk parking brakes at the front.

10. Hydraulic System

Independent hydraulic systems are used for travel drive, conveyor belts, cooler fan drive, water spray system and setting cylinders. The hydraulic pump are driven by the diesel engine via a splitter gearbox.

11. Electric System

The voltage of electric system is 24 V with starter, 3-phase alternator and two 12 V batteries, as well as socket outlets for lamps.

12. Water System

A hydraulically operated water supply system cools the point-attack cutting tools during milling, thus considerably extending their service life. The water pressure and quantity can be adjusted from operator's platform. The spray nozzles are easily removed for cleaning.

In addition, both the primary and the discharge conveyor can be sprinkled separately at several points. The continuous sprinkling of the drum and the conveyor system effectively prevents the development of dust.

New Words

1. carriageway[ˈkærɪdʒweɪ]	n.	车道,马道
2. reclaim[rɪˈkleɪm]	vt.	要求归还,开垦
3. manoeuvre[məˈnuːvə]	vt.	调度,超重
4. slew[sluː]	vi.	旋转,猛地转向
5. patent[ˈpæt(ə)nt; ˈpeɪt(ə)nt]	vi.	获得专利
	n.	取得专利权

6. established [i'stæbliʃt]		adj.	得到承认的
7. joystick [dʒɔistik]		n.	操纵杆
8. nuisance ['njuːs(ə)ns]		n.	令人讨厌的人,事情或状况
9. polyurethane [ˌpɔli'juərəθein]		n.	聚氨酯
10. inclement [in'klem(ə)nt]		adj.	险恶的,严酷的
11. prevail [pri'veil]		vt. vi.	盛行,胜过
		adj.	主要的,占优势的
12. scan [skæn]		vi.	扫描
13. dimension [di'menʃ(ə)n ; dai-]		n.	尺寸;(复)面积,体积
14. sprinkle ['spriŋk(ə)l]		vt.	洒,淋
		n.	洒,淋,小雨

Phrases and Expressions

1. asphalt pavement — 沥青混凝土路面
2. low-bed trailer — 低车架拖车
3. retaining bolt — 固定螺栓
4. up-milling direction — (刀具)从下向上铣削
5. milling drum — 铣刨滚筒
6. V-ribbed steep-incline belt — 加筋三角提升皮带
7. discharge conveyor — 卸料输送器
8. rotary transducer — 旋转传感器
9. differential lock — 差速锁
10. self-locking hydrostatic transmission — 自锁式液压变速器

Notes

①The scraper blade opens hydraulically (can be slewed by 100°), thus providing good access to the milling drum for replacement of cutting tools.

刮板由液压装置开启(可以转动100°),为更换铣刨鼓上的切削刀具提供了通道。

②The discharge conveyor can load trucks from a great height, its height is adjustable and can be slewed to both sides, thus always allowing an optimum adaptation to the conditions prevailing on site.

卸料输送器可以从很高的位置给载货车卸料,卸料的高度是可调的且可向两边转动,因此,总是能最佳的适应工地条件。

③The reference planes can be scanned by various methods, for instance, by a wire-rope sensor at the side plates, an ultrasonic sensor on the existing road surface, a grade-line in combination with rotary transducers, or a plane formed by lasers.

参考平面可以通过不同的方法进行扫描,例如,可以用侧板上的钢丝绳传感器扫描、旧路面上的超声波传感器扫描,对道路纵坡线使用角位移传感器扫描或用激光扫描。

Exercise One

Put the following expressions into Chinese:

1. asphalt pavement
2. cold milling machine
3. milling drum
4. V-ribbed steep-incline belt
5. discharge conveyor
6. differential lock

Exercise Two

Put the following sentences into Chinese:

1. Hot milling machines with heating devices are usually used on operation of hot recycling road.

2. The cold milling machines are used to remove the defective paving to the required thickness.

3. The milling machines can be classified as front loading up cutting and rear loading down cutting types.

4. The discharge conveyor is covered to prevent clouds of dust from being blown away by the wind and causing a nuisance.

Lesson 11

Hot Recycling Machines

The hot recycling method is used to repair bituminous bound surface courses which produces cracks, ruts, deformation, aging and abrasive by replastifying the pavement and admixing binders and additional mixing material.

Thehot recycling methods include remixing (Fig. 2-11-1) and repaving recycling (Fig. 2-11-2) constructions. Remixing method is to remix reclaimed material with an additional meterial and then to repave the mixture. Repaving method is to profile the existing pavement and then to cover it with a new layer.

Repaving method is commonly used to repair and renovate construction of damaged pavement or to upgrade and improve aging pavement, and remixing method permits restoration of the service properties of the pavement, both making complete reuse of the existing road materials.

The hot recycling machines include on-site mixer and at-plant mixer.

Hot recycling machine is illustrated in Fig. 2-11-1 for rehabilitating large areas of bituminous bound carriageways. Infinitely variable working width is for taking up and placing the material mixed with binder and virgin asphalt. It can operate continuously to on-site heat, scarify, mix, pave, compact. It includes recycling heater, recycling machine (repaving machine, remixer), feeding device of admixture, scarifier, mixer, remixture paving device, feeding device of additive material, screed and grade and slope control of screed.

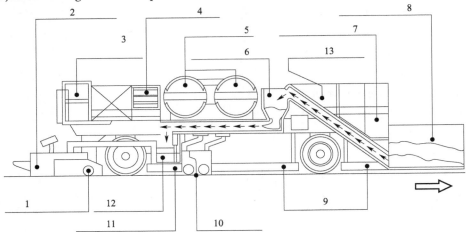

Fig. 2-11-1 Hot recycling machine(remixing)

1-spreader auger;2-variable screed;3-operating stand;4-engine;5-gas tanks;6-batching bin;7-heated binder tank;8-receiving hopper;9-infra-red heaters;10-variable scarifiers;11-infra-red heaters;12-pugmill;13-fuel tank

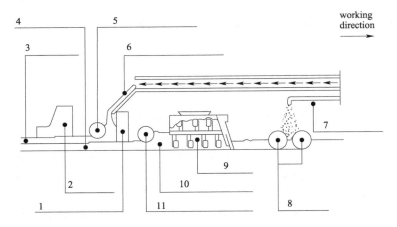

Fig. 2-11-2 Hot recycling machine (repaving)

1-1st screed;2-2nd screed;3-course of vingin mix;4-remixed surface course;5-2nd spreader auger;6-feeding of virgin mix;7-bitumen injection;8-scarifiers;9-mixer;10-remixed material;11-1st spreader auger

1. Heating Device

The existing pavement is softened by infrared heaters. The base material is also heated in this way before the mixture is laid (hot-in-hot placing). The energy source used is propane gas, which is vaporized and burnt in gaseous form.

Gas tanks: double tank for liquid gas, with level indicator.

Vaporizer: Two gas-operated vaporizers with thermostatic control.

Heating elements: All-metal heaters, can be swung out individually to suit the working width.

Heating capacity: Control valves for distributing the heat output in the individual heaters (setting the working pressure).

2. Admixture

The admixture specified by laboratory is loaded into the remixer in batches from trucks and is continuously mixed into the reclaimed material in the appropriate proportions.

Receiving hopper: Capacity is specified and with hydraulically tilting side walls.

Pushing rollers push the truck during the unloading process.

Inclined conveyor: Robust scraper conveyor with wear-resisting drag slats and roller chains are hydrostatically driven. The conveyor tunnel is heated so that the mix does not cool down.

Batch bin: Supply container for the admixture with push-in metering devices. Measuring is infinitely adjustable.

Chassis conveyor: Robust scraper conveyor with wear-resisting drag slats and roller chains are hydrostatically driven and infinitely adjustable. The conveyor tunnel is heated. The admixture is either fed into the mixer (flap opened hydraulically) or in front of the screed.

Automatic quantity control: The pre-selected quantity to be added (per m^2) is continuously monitored by controlling the chassis conveyor in accordance with the forward advance speed of the machine.

3. Scarifier Unit

Scarifier unit consists of two scarifier shafts and leveling blades which auger the material inwards and two shafts with leveling blade for augering the material to the mixer. The softened pavement is loosened by the rotating scarifier shafts with cutter teeth and augered into the machine with the help of the leveling blades. The working width can be infinitely adjusted via the hydraulic system. Suspension: The scarifier unit is suspended on rods positioned on both sides and is positioned by two hydraulic cylinders in accordance with the desired loosing depth.

Loosing depth: An automatic device keeps the selected loosening depth constant.

4. Pugmill Mixer

A compulsory pugmill mixer thoroughly mixes the reclaimed material with the admixture until it forms a homogeneous mass.

Mixer: Horizontal twin-shaft compulsory pugmill mix with high-strength lining are hydrostatically driven. The mixer is heated.

5. Screed and Level Control

②The prepared material is discharged from the mixer in a window and accurately placed to profile by the infinitely variable screed.

Compacting screed: Tampering and vibrating screed are hydraulically operated for high pre-compaction. Surface profile can be adjustment.

An additional screed is for precise paving of the existing asphalt pavement which has been sprayed with bitumen prior to overly.

Grade and slope control for screed consist of two grade control and one slope control. The reference levels are sensed either on the side plates or on taut wires. Hydrostatically driven brushes for cleaning the joints adjacent to the existing pavement located in front of the screed.

6. Bitumen

Bitumen tank: It can be fed with liquid bitumen or with bitumen blocks. It is equipped with a thermostatically controlled heating oil heater.

Bitumen pump: Quantity to be added can be infinitely adjusted via the pump speed.

Metering control: The preselected quantity to be added (per m^2) is continuously monitored by an automatic control device by controlling the binder pump in accordance with the forward speed of the machine.

Spraying device: It includes heated spraying tube with nozzles. The bitumen is sprayed in front of the two shaft pugmill mixer and thoroughly mixed in the mixer.

New Words

1. replastify [ri'plæstifai]   vt.   再塑,重塑,整形
2. renovate ['renəveit]   vt.   更新,修复,革新,刷新
3. upgrade [ʌp'greid]   n.   升级,上升,上坡
                                      vt.   使升级,提升,改良品种
4. pugmill [pʌgmil]   n.   拌泥机
5. restoration [restə'reiʃ(ə)n]   n.   回复,恢复,复原
6. virgin ['vəːdʒin]   adj.   纯洁的,原本的,未开发的,首次的,创始的
7. vaporizer ['veipəraizə]   n.   蒸发器;汽化器
8. scarifier ['skeərifaiə, 'skæ-]   n.   松土机,翻土机
9. rehabilitate [riːə'biliteit]   vt., vi.   恢复;修复,更新
10. propane ['prəupein]   n.   丙烷
11. pendulum ['pendjuləm]   n.   钟摆,摇锤
12. reclaim [riːklem]   v.   回收利用
13. compulsory [kəm'pʌlsəri]   adj.   被迫的,被强制的;义务的
14. taut [tɔːt]   adj.   拉紧的,紧张的

Phrases and Expressions

1. inclined conveyor   斜置传送带
2. loosening depth   铣刨深度
3. bitumen spray system   沥青喷射系统
4. grade and slope control   坡度控制
5. vibrating screed   振动式熨平板
6. infrared heater   红外线加热器

Notes

①The unit consist of two scarifier shafts and leveling blades which auger the material inwards and two shaft with leveling blade for augering the material to the mixer.

Part 2　Engineering and Construction Machineries

翻松装置由两个旋转滚筒式翻松装置和刮刀组成,刮刀将翻松的材料向内聚集,且将材料螺旋输送到搅拌器内。

②The prepared material is discharged from the mixer in a window and accurately placed to profile by the infinitely variable screed.

搅拌好的混合料从搅拌器的卸料口卸出,并由无级变宽的熨平板精确的摊铺形成路面轮廓。

Exercise One

Put the following expressions into Chinese:

1. inclined conveyor
2. loosening depth
3. bitumen spray system
4. grade and slope control
5. vibrating screed
6. infrared heater

Exercise Two

Put the following sentences into Chinese:

1. Remixing method is to mix reclaimed material with an additive and relaying the mixture.
2. Repaving method is to profile the existing pavement and covering it with a new layer.
3. Grade and slope control for screed consist of two grade control and one slope control sensors.
4. The reference levels are sensed either on the side plates or on taut wires.

Unit 2 Reference Translations and Answers

第一课 推 土 机

推土机是一种土方机械,由拖拉机和安装在履带和轮式拖拉机上的铲刀组成(图2-1-1)。[①]在铺设路基中,推土机用来剥离草皮覆盖层、清除灌木、树桩和小树;挖土并把土从挖土处运到某一堆土处;筑路堤、推土至弃土堆;在山坡上筑路基时推土、平整挖土机堆积的土堆、填平沟、渠和坑洼、运淤泥到桥墩处。

图 2-1-1 履带式推土机
1-推土铲;2-液压系统;3-履带式拖拉机;4-松土器

推土机还可以做一些辅助工作,如:为铲运机和自卸车清理线路;当铲运车满载时助推;可以安装动力铲或液压装置工作。

在冬季推土机可以在机动车道、机场和市政广场铲雪。

在仓库和施工现场推土机可以运输、堆积,装载碎石、砂子、砾石和其他的松散材料并开发小型的砾石和砂子采石场,把石料输送到拌和机的供料装置中。

推土机最适合于100m距离以内推运土壤和松散的建筑材料。推土机可以筑路堤可以挖方,高度和深度可达3m。

推土机分类如下:

(1)按刀铲的安装分:直推式:其铲刀垂直于拖拉机的纵轴线(铲刀处于这个位置推土机把挖土向前推);万能式:其铲刀或和拖拉机纵轴线垂直或成一定的角度(通常60°),向任何一方推土。铲刀也可以在垂直面内斜置±5°~8°,可以将土向前推也可以向一侧推,填沟渠、筑路堤。

（2）按控制系统分：钢索操纵式：其铲刀的升降是靠安装在拖拉机上的绞盘带动钢索来实现；液压操纵式：其铲刀由油泵通过压力把工作液提供给液压缸来操纵。液压式推土机铲刀强制入土，所以在全球应用越来越广泛。

（3）按牵引车类型分：履带式、轮式和专用汽船式拖拉机。

履带式推土机的优点如下：

（1）能传递更大的牵引力，尤其是在松软的路基上工作，如松软的、泥泞的土壤；
（2）能在泥泞的土壤上行驶；
（3）能在岩石层上工作，而轮胎式推土机可能就会严重损坏；
（4）能在崎岖的路面上行驶，但可能减少经济运距；
（5）浮力大，因为履带的平均压力小；
（6）更高的通用性。

轮式推土机的优点如下：

（1）从一个工地到另一个工地的运行速度更快；
（2）推土机到工地省去了运送设备；
（3）更大的功率输出，尤其是大运距的情况下；
（4）工作不易疲劳；
（5）在路面上行驶时不会破坏路面。

Key to the Exercises

Exercise One
1. 安装在履带或轮式拖拉机上的悬挂装置
2. 机动车道、机场和市政广场
3. 穿过泥泞土壤的能力
4. 桥墩
5. 路基
6. 履带式推土机

Exercise Two
1. 直推式推土机的刀铲与拖拉机的纵轴线垂直。
2. 液压式推土机在国际上得到了更广泛的认识。
3. 液压式推土机的刀铲与钢索操纵式推土机的刀铲入土速度相同。
4. 履带推土机比轮式推土机更先进。

第二课　铲　运　机

铲运机在土方施工中一层一层铲土、运输并堆放土壤，即将土壤推到卸土场并进行刮平。①当在新开挖的土壤上行驶时，铲运机还可对其进行部分压实。铲运机广泛应用于筑路、工业建设、水利工程、矿山的剥离及土方工程中。

铲运机可适用于包括黑土、沙土以及黏土的各种土壤的施工,而不适用湿度较大的黑土或黏土,因为它们会黏结在斗壁上堵塞铲斗。因为车轮易沉陷,铲运机最不适合在沼泽地段工作。

松散的沙土不能很好地装满铲斗,并且卸载困难。实践证明,中等湿度的沙土和垆姆土能够完全装满铲斗,此时效率最高。铲运机不能在含石量大的土壤中工作,过大的土块应在铲运前破碎。目前,在许多国家已经研制了刮板送料器和提升机构帮助更满的装料、装载更均匀、更有效地利用拖拉机的牵引力。

铲运机可以如下分类:
(1) 按铲斗的容量分为:$1.5m^3$、$3.0m^3$、$6.0m^3$、$10.0m^3$、15.0 和 $25.0m^3$ 等;
(2) 按行走机构分为:拖式、半拖式和自行式三种;
(3) 按卸载方式可分为:自由式、半强制式和强制式;
(4) 按控制系统可分为:液压控制和钢索控制。

当铲运机必须铲装坚硬的土壤时,用一台牵引车或一台两轮牵引车,动力不足以切削土壤并装满铲斗,此时应用一台履带式牵引车或四轮顶推牵引车帮助主机前进。

铲运机作为即考虑最佳装料又考虑最佳运料的组合机械,不可能在装料和运料两方面都强于其他机械。强力挖掘机、拉铲挖掘机、带式装载机在装料方面会超过铲运机,而载货车在运料方面也会超过铲运机,特别是在运距长、养护良好的路况下更能体现出来。

根据助推牵引车的类型铲运机还可以分为两种:履带助推式牵引车和轮式助推牵引车。助推牵引车顶在铲运机的后保险杠上,并和拖拉牵引车一起产生必要的作用力装斗。鉴于切土和装料过程占整个工作循环很短的一段时间,一台助推牵引车可以为几台铲运机服务。

②运距较短时,履带牵引车助推轮胎、自行装载式铲运机可以更经济地运土。另外,履带牵引车还具有其他的优势:如,装料时具有较高的牵引力,同时具有良好的附着力;在较差的路况下短距离能有效地运土。随着运距的增加,与轮胎牵引车相比,履带牵引车会处于劣势。

除非装载非常困难,履带牵引车给铲运机助铲不需要推土机的帮助。如果几台铲运机联合工作,就需要推土机助铲,此时动力输出得增加足以证明推土机使用的正确性。

运距较长时,轮式牵引车助推自行装载铲运机与履带式牵引车助推相比,较高的行驶速度使运土更经济。尽管轮式牵引车在铲运机装料时不会产生更大的牵引力,但运距足够长时行驶速度快又弥补了它的缺点。

铲运机装料是通过放下铲斗的前端,直到沿铲宽度方向布置的铲刀切入泥土,与此同时,提起前挡板形成一个开口让土壤流进铲斗内。

当铲运机向前拉时,带状土被切进入铲斗,这个过程一直持续到铲斗被填满(不能再装更多),此时,抬起铲刀,放下前挡板以防止在运输过程中土壤的溢出。

卸料过程是将铲刀降低到填方上面预定的高度,提起前挡板,借助于安装在铲斗后部的可移动式卸料器强制的使土从铲刀与前挡板之间卸出。铲运机工作过程如图 2-2-1 所示。

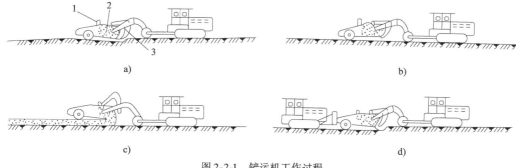

图 2-2-1 铲运机工作过程
a)装载过程;b)运输工程;c)卸载工程;d)推土机助推
1-卸料板;2-铲斗;3-斗门

Key to the Exercises

Exercise One
1. 最佳装载(量)和最佳运距
2. 维护良好的运输线路
3. 履带拖拉机
4. 土方结构
5. 轮胎式自行装载铲运机
6. 筑路、工业和水利工程

Exercise Two
1. 履带拖拉机拖曳的铲运机行驶速度非常慢。
2. 当运距增加时,履带拖拉机与轮式拖拉机相比缺点是行驶速度低。
3. 当装土时铲运机更适合中等适度的砂土和垆坶土。
4. 当铲运机向前推时,带状土被铲进斗中。

第三课 装 载 机

在复杂的机械化施工中,机械化装载和卸载是非常重要的。为此,各种装载机被使用且分类如下:

按用途分:处置块状或松散的物料;

按工作过程的特征分:间歇工作单斗式和连续工作多斗式;

按行走装置分:履带式和轮胎式;

按发动机类型分:电源驱动的电机,内燃机和柴油机—电力传动装置;

卸载装置处理铁道运送的物料,并且可以改装来满足载荷的自然需求。

特殊的卸载装置用来处理砂、砾石和水泥。其他的物料可以通过起重机械(抓斗起重机等)的帮助卸载。

根据操纵单斗装载机的方式不同,单斗装载机还可以分为前卸式(料斗向前倾翻);回转式(部分回转,料斗向侧面转动、卸料);后卸式;组合式。

实际上,①前卸式和回转式装载机都可以更换工作装置,比如,大、小料斗,抓斗,叉子,钩子等,因此,它的应用范围更大。这些装载机用来装载或卸载松散物料,如,土壤、卵石、煤和砂等。它还用在矿井诸如拔桩的工作。当使用辅助装置时,它还可以代替推土机进行摊铺、清理工作表面、整平土堆。这种类型的装载机是一种理想的工作装置,广泛用在矿井、筑路、码头和森林施工中,还用在其他的基础施工工作中。

当今的施工中,广泛使用铰接车架装载机,这样大大增加了转向性能。如果装载或卸载时没有严格的对正需求,它们还具有转弯半径小、机动灵活的优点。即使在狭窄区域工作,也可得到满意的效果。

传动系统使用了液力变矩器,它使装载机更有效、并且对发动机和变速器的工作部件提供保护作用。

单斗装载机的工作装置包括提升臂、提升油缸、倾翻油缸、倾翻推杆、倾翻联动装置和料斗,如图 2-3-1 所示,两个提升油缸控制斗臂,使料斗提升或降落。两个倾翻油缸伸出或缩回使料斗转动,因此物料可以装在料斗中也可以从料斗中卸出。

目前,由容积排量式液压泵驱动工作装置的前卸式装载机应用最为广泛。

多斗装载机的功率输出比相同标定功率的单斗装载机高 40%~60%。多斗装载机适用于砖厂、混凝土预制件厂及装卸量大的火车站,同样也适用于装载松散物料。另外,它还适用于对松散物料的颗粒分级,为此专门配备了专用的振动筛。

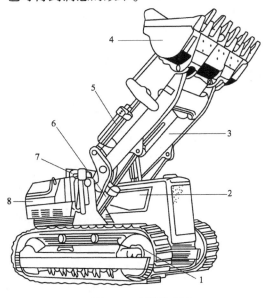

图 2-3-1 履带装载机
1-履带行驶机构;2-发动机;3-提升臂;4-料斗;5-倾翻油缸;6-提升油缸;7-驾驶室;8-油箱

多斗装载机可以为侧面的平板车卸料,其效率很高。它也用于混凝土预制件的生产线中,也可用于筑路施工中。在筑路施工中,主要用途是将砂和砾石装入烘干筒和搅拌机中。该装载机的工作机构是由左右两个螺旋杆组成的螺旋给料器组成,它们分别安装在斗式提升机的两侧。当给料器转动时,物料被送到料斗中,使装料容易。刮板固定在螺旋给料器的下方。多斗装载机的提升机把物料卸在传送带上,再由传送带将物料运送到运输装置中。

有些装载机用漏斗或溜槽把物料送进运输设备中。

Key to the Exercises

Exercise One

1. 装载和卸载机械化
2. 铰接车架装载机
3. 混凝土预制件厂
4. 多斗装载机

5. 间歇操作单斗装载机
6. 液力变矩器

Exercise Two
1. 装载机可以应用在预制件生产线和筑路施工中。
2. 实践证明挖掘机在矿山采石和储料施工中装载量不如装载机。
3. 履带装载机通过性好且可以产生更大的牵引力。
4. 有些装载机通过料斗和料槽把料卸到运输设备中。

第四课 挖 掘 机

①强力挖掘机是一种以铲斗作为工作装置的挖掘机,它挖出泥土运送一小段距离后,把它堆成土堆或卸在运输设备上。

回转挖掘机用于各种工程施工中,如挖铺设排水管的道沟、浇灌土地的水渠,开采有价值的矿产,在采矿中做剥离工作,露天开采煤和矿石。回转挖掘机可以按作用、斗容量、使用范围、动力源、行走机构和控制系统分类。

回转挖掘机按斗容量、作用可以分为以下几种:

（1）组合式挖掘机:斗容量 0.15~4m³,与各种不同的替换装置一起工作,用于工程施工与辅助工作,并在小型采石场用来挖掘建筑材料。

（2）采石挖掘机:斗容量 4~8m³,在采石场挖掘原料,在爆破后处理岩石,挖掘黏质土。

（3）剥离挖掘机:斗容量 4~10m³,在露天开采矿石中用于清理盖岩层并堆积成堆。

（4）移动拉铲式挖掘机:斗容量 4~50m³,在水利工程中用来挖掘软土深坑。

回转式挖掘机设计成可以和各种不同的替换装置(工作装置或前部装置)一起使用,如:正铲(图2-4-1a);反铲(反向铲)(图2-4-1b);拉铲(图2-4-1c);抓斗(图2-4-1d);悬臂式起重机(图2-4-1e);塔式起重机、打桩机(图2-4-1f)和冻土松土器,其他设备包括刨刀、填沟器及重力振捣器。

回转挖掘机由下列主要部件组成:工作(前部)装置(铲斗、动臂、斗杆和液压油缸);安装动力装置的转台、动力传动系统和液压系统;承受转台和各种工作机构和各种设备重量的车架;支撑车架的行走装置(履带或轮式)。

单斗挖掘机的工作循环(图2-4-2):挖掘、从工作面提升铲斗 B;使斗杆向前 C;铲斗从工作面拉回 D;旋转动臂到卸料处 E 为卸料调节铲斗位置 F;打开斗底门卸料、转动铲斗返回到工作面同时设定在最初的工作位置 A。

挖掘机的动力可能是柴油机、电动机(外部电源驱动)或柴油发电机组。

液压控制工作装置的挖掘机是指刚性连接工作装置的挖掘机,它与挠性连接工作装置的挖掘机是不同的。

挖掘机工作过程中,铲斗刀刃上所承受的总作用力通常可以分解为切向和平行(斗杆的)挖掘力。就铲斗为刚性连接的挖掘机而言,挖掘力(也就是从土中提升铲刀)通常由挖掘机的总重量来保证。而铲斗为挠性连接的挖掘机,其挖掘力仅靠工作装置的重量来保证。刚性连接(液压)挖掘机的这个特点可以提高挖掘量,尤其是在挖掘深度较大时更为突出。

液压挖掘机在结构、工艺及使用方面与挠性连接式挖掘机相比有许多的优势。

(1) 结构方面：

①可无级调节行走速度和牵引力,便于倒车；

②可用多发动机驱动方式而不增加机器的总重量,也不必安装那些需要精心保养的传动装置。

(2) 使用方面：在工作状态下,比较容易控制操作机构和工作装置（速度）,易于使旋转运动变成往复运动。

(3) 技术方面：当斗杆相对于动臂固定不动时,只要反向转动铲斗就能挖掘,使其在狭窄的地方,也可进行挖掘作业。当铲斗容量相同时,液压挖掘机的总体尺寸要比挠件连接式挖掘机小很多。

(4) 经济方面：

①由于可以替换多种附属装置,刚性连接式挖掘机可以大量减少劳动力,提高生产率；

②易于利用标准件组装成多种机械,从而使制造和维修简化。

液压挖掘机便于实现自动化,因此,可改善挖掘机操作者的工作条件,提高时间利用率。

单斗液压挖掘机的工作装置、行走装置、回转机构和其他装置都是通过液压传动装置和液压马达驱动的。

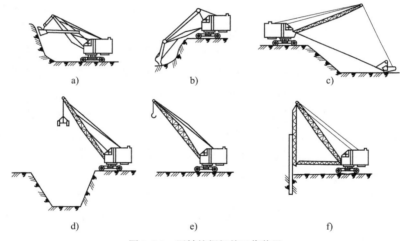

图 2-4-1　回转挖掘机的工作装置

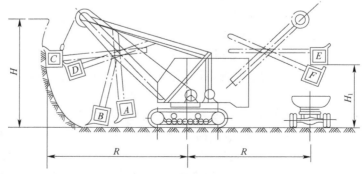

图 2-4-2　回转挖掘机的工作循环

Key to the Exercises

Exercise One
1. 液压挖掘机的工作装置
2. 挖掘机的结构、液压系统和挖掘的几何尺寸
3. 重力和惯性力
4. 拉铲挖掘机起重臂
5. 单斗挖掘机的工作容量
6. 回转式挖掘机

Exercise Two
1. 斗容量为 4～10m³ 的挖掘机在开采露天矿施工中用来清理覆盖层并堆成堆。
2. 尤其是挖掘深度达到一定深度时,刚性悬挂装置的结构特点是增加输出功率。
3. 回转式挖掘机包括工作装置、转台和行驶装置。
4. 刚性连接式挖掘机便于实现自动化,因此,可改善挖掘机操作者的工作条件,提高时间利用率。

第五课　平　地　机

自行式平地机在公路施工中用于挖边沟,为填方和挖方的表面、边端整形达到所需要的坡度。

自行式平地机可以以水平面为基准做铺设路基、整平及辅助工作,为挖方和路堤筑台、整形。在冬季,自行式平地机还可以清除道路积雪。

自行式平地机是机动性非常高的机械,它的刮刀在水平面或纵垂面都可以在不同的角度设定,还可以从机身侧面伸出。

自行式平地机按车轮数目分为四轮和六轮两类,以转向轮对数×驱动轮对数×车轮总对数表示,主要有 1×1×2;2×2×2;1×2×3;1×3×3;3×2×3;3×3×3 六种类型。转向轮越多,转向半径越小;驱动轮越多,牵引力越大。

F155 型平地机是一种施工机械,如图 2-5-1 所示。

发动机是 5 缸、4 冲程、直喷、风冷发动机,汽缸直列。发动机的动力传递给与发动机相连的变矩器、带分动箱的 4 挡动力换挡变速器,再通过主传动轴传递给后桥,然后通过平衡串联传动箱传递给车轮。

前轮液压转向、底盘铰接使平地机具有极高的机动性能和最小的转弯半径。因为大功率、多功能的工作装置,F155 型平地机可以更平稳、更经济地工作。

前、后车架是刚性的、焊接在一起的钢制结构,后车架上安装发动机、变速器、平衡箱,液压油和燃油箱、制动系、驾驶室。前车架上安装刮刀、刮刀环形刀架、牵引架、支座和液压油缸。前、后车架铰接连接。

安装在中央的刮刀是平地机的主要工作装置。因为切削角度和刀板可以调整,所以刀刃的设置可以保证最佳的刮平和切削质量。

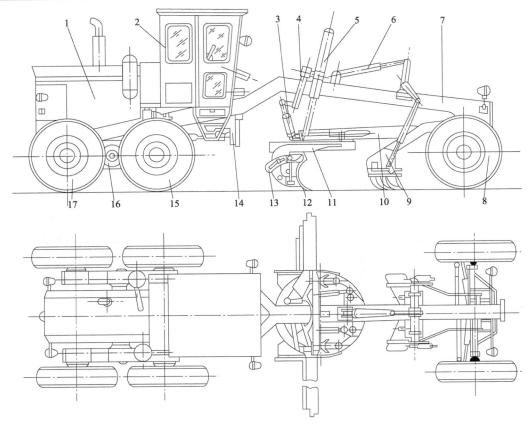

图2-5-1 自行式平地机

1-发动机;2-驾驶室;3-牵引架引出油缸;4-摆架机构;5-升降油缸;6-松土器收放油缸;7-车架;8-前轮;9-松土器;10-牵引架;11-回转圈;12-刮刀;13-角位器;14-传动系统;15-中轮;16-平衡箱;17-后轮

　　平地机刮刀安装在环形刀架上,通过液压马达驱动的蜗轮、蜗杆装置使环形刀架转动,环形刀架和牵引架相连,①牵引架通过滚珠回转接头安装在平地机机架的前部。六齿松土器装在平地机刮刀的后面,可以向前或向后翻土。刮刀的工作(如升、降)是通过液压油缸控制,液压油缸由液压阀控制板上的手动阀控制。

　　刮刀辅助工作装置用来做粗略的平地工作,后部的松土器为硬土松土,松土器上安装了松土齿耙并由液压缸控制,液压油缸由液压阀控制板上的手动阀控制。松土器板梁(宽度)可以达到平地机的行驶宽度。

Key to the Exercises

Exercise One

1. 自行式平地机
2. 直喷式发动机
3. 铰接车架
4. 液压缸
5. 松土器

6. 最佳铲、挖质量

Exercise Two

1. 自行式平地机在公路施工中用于挖边沟,为填方和挖方的表面、边端整形达到所需要的坡度。

2. 前、后车架是刚性的焊接在一起的钢制车架。

3. 平地机刮刀安装在环形刀架上,通过液压马达驱动。

4. 自行式平地机是一种机动灵活的机械,铲刀在水平面和纵垂面都可以设置,还可以向侧面伸出。

第六课 沥青混凝土摊铺机

沥青摊铺机即沥青成型机,是摊铺沥青拌和设备搅拌的沥青混合料的专用机械。它可以快速、均匀地将拌和好的沥青混合料摊铺在路基上,并且进行压实和整平。

沥青摊铺机的工作过程如下:倒退中的自卸车的后轮顶在沥青摊铺机的推滚 12 上,并将沥青混合料卸到的料斗 13 中。由独立驱动的左、右两个刮板输送器 10 将沥青混合料输送到螺旋输送器 8 处。沥青材料进一步向路的两侧摊开。振捣棒 7 进行初步压实。包括振动器、加热装置、调拱装置和熨平板的熨平装置 6 使沥青混合料形成确定的形状和厚度。摊铺机的主要构件如图 2-6-1 所示。

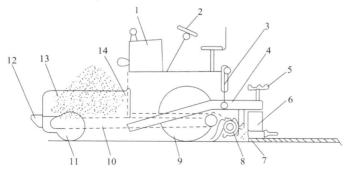

图 2-6-1 摊铺机

1-发动机;2-方向盘;3-升降油缸;4-侧臂;5-厚度控制器;6-熨平装置;7-振捣棒;8-螺旋布料器;9-驱动轮;10-刮板输送器;11-前轮;12-推滚;13-料斗;14-供料斗门

1. 推滚

当处于怠速的运料自卸车往摊铺机料斗中卸料时,推滚使它与摊铺机同步前行直到卸完料自卸车驶离。

2. 熨平板

熨平板是摊铺机最重要的部分,①它完成摊铺工序并形成铺层的最终状态,达到必要的整平度和初压实厚度。

熨平板可配置振捣棒也可配置振动器。振捣棒的作用是将混合料塞进熨平板的导料缘下,同时,对混合料进行一定程度的压实。

当摊铺机主要工作是摊铺碎石和干硬的混凝土时,装有振动器的熨平板能非常有效地摊铺这些材料,并形成均匀的、平整的表面。

熨平板包括两个主要部分,构成了基本熨平板和加长段,即可以加长到所需的摊铺宽度。每一段(包括延伸段)都具有坚固的结构和精确的加工,以保证精确匹配并形成一个刚性的、非挠性的整体构件。熨平板的长度应保证每一段都具有平顺的(摊铺)表面。为了有利于混合料在振动熨平板下流动,熨平板底板具有一定的仰角并配合一个可调整的整平板。熨平板在铺层上可以形成3%的路拱。熨平板装置还配有火花塞和燃油加热系统,使熨平板达到并保持工作温度。有些摊铺机还配备电加热装置或丙烷燃烧器。

3. 料斗

安装在摊铺机前端的料斗接收自卸车卸下的混合料,且要容纳足够的混合料确保自卸车换车时摊铺机能正常摊铺,这样可以保证施工的连续性且可以使铺层表面光滑。

料斗底部配有两个刮板输送器,并且在左、右两端还配有两个可移动的斗壁,斗壁在油缸的作用下可以升降,这样可以使斗中所有混合料都送到刮板输送器上。

4. 供料斗门

在车架前端安装了左、右两个供料斗门,它们用来控制料斗给摊铺室提供的混合料的量。它们分别由两个油缸控制。

5. 刮板输送器

摊铺机使用两排刮板输送器,它们分别由气阀控制。当拉杆通过气阀向后推时,供料装置供料;拉杆向前推时,停止供料。

6. 螺旋布料器

螺旋布料器将混合料均匀地分布在熨平板导料缘的前边,它的结构如图 2-6-2 所示。

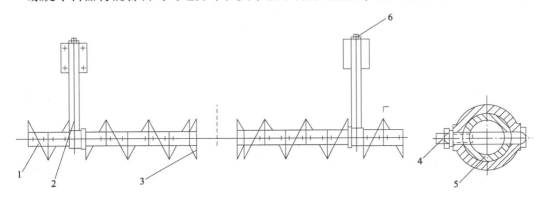

图 2-6-2　螺旋布料器

1-节距;2-支撑臂;3-叶片;4-连接螺栓;5-连接套筒;6-后板

通过支撑(臂)与车架相连的螺旋布料器分左、右两部分,②与用来控制相应的刮板输送器的同一气阀控制。

螺旋布料器分为两部分,叶片以一定的节距分段安装并通过套筒和螺栓把它们连接起来。

左侧螺旋布料器左旋,右侧布料器送器右旋。

在左、右螺旋布料器内侧的端头,装有中间反向叶片保证向中间供料。

为了帮助操作人员控制铺层平整精度,摊铺机配有一个完整的自动找平控制系统,任何一种组合方式都依赖于电信号控制的两侧大臂的牵引点。两侧大臂都是由独立的方法(油缸)控制,或者是纵向(基准线)传感器,或者是横坡传感器。传感器可以根据基准线接收信号,基准线可以是预先架设的张紧的钢丝绳;由路缘石、隔离墩或相邻路面形成的固体参照物;或者可以在6~12m距离内的平均路面段内使用的浮动梁。自动找平控制系统由这些方法中的任意两种组合而成,用来控制两侧大臂的牵引点。

通常,沥青摊铺机由柴油机提供动力,柴油机采用柔性安装,以防止振动传递给摊铺机的其他部件。分别给牵引、两个输送系统、振捣和/或振动器以及各种油缸独立提供液压油的四个油泵安装在与发动机直接相连的分动箱上。

沥青摊铺机可以分为履带式和轮式。

Key to the Exercises

Exercise One
1. 铺层质量
2. 熨平板和延伸段
3. 沥青混凝土摊铺机
4. 混合料
5. 料斗
6. 举升油缸

Exercise Two
1. 从料斗送料带接料后,链传动的螺旋输送器把料均匀的、沿着整个熨平板的宽度摊铺在导料缘的前方。
2. 混合料从料斗底部通过一对斗门进入到螺旋输送器箱内。
3. 操作者很容易观察到整个的螺旋输送器的槽、料斗还有供料自卸车以及运行方向指示器。
4. 沥青混合料摊铺机由弹性安装的柴油机驱动,防止振动传递到摊铺机的其他部件上。

第七课 水泥混凝土摊铺机

水泥混凝土广泛用在筑路施工中,也广泛用在水泥混凝土路面机械中。这些机械可以分为两种:轨道式摊铺机和滑膜式摊铺机。

现代混凝土摊铺机大多数是滑膜式摊铺机,因为它的生产率高、结构紧凑。由于滑膜摊铺机的工作精度接近完美,所以生产率高且表面偏差可以控制在最小的范围内。近几年来,滑膜摊铺机的技术已经发展到与任何一种摊铺方法相比每平方米成本更低、摊铺质量更好的程度。

针对大负荷的施工,滑膜摊铺机配备了摊铺装置、振动块和强制(液压)驱动装置,而且以完美的行走直线、高生产率摊铺矩形、垂直侧边(立方体)的面板。因为独特的行走动力分配系统,滑膜摊铺机是不会出现拖延工作或混合料堵塞现象的。

滑膜摊铺机可以在钢筋上形成面板,也可以没有钢筋。它可以分配、计量倾卸到它前面路段里的混合料量,调整、抹光由布料器最初形成的面板。

SF-350 四履带滑膜摊铺机如图 2-7-1 所示。

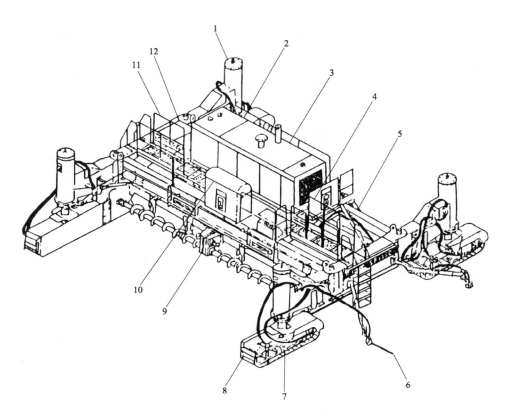

图 2-7-1　SF-350 四履带滑膜摊铺机

1-支腿总成;2-油箱;3-动力系统;4-人行通道;5-伸缩机架;6-自动找平系统;7-转向系统;8-行走装置;9-摊铺装置;10-操作控制台;11-固定机架;12-边架

在工作过程中,倾卸到摊铺机前端的水泥混凝土通过螺旋布料器沿着机器前端均匀地摊开,在螺旋布料器后面的可调虚方控制器只允许设定量的混凝土通过虚控制器到达浮动盘前面的振动器和振捣棒处。①振动器提浆以使混凝土密实且容易成型;而振捣棒则将大集料挤到面层下方以便成型盘形成均匀表面层。成型盘后面的浮动盘抹光面板的上表面。成型盘两端的侧模版形成面板的边缘,而浮盘两端的磨边器抹平面板的侧边。

摊铺机的功用靠五个独立的液压系统来完成,它们分别是行走驱动系统、螺旋布料器驱动系统、振动器驱动系统、振捣棒驱动系统和辅助液压系统。

所有系统的动力都由(主)动力系统提供,机器的行进靠四个履带总成完成。

单车道滑膜摊铺机与双车道滑膜摊铺机除了具有相同的优点外,还具有在更狭窄区域能有效工作的优点。在现有道路加宽的工程中,在坡道、中央隔离带、交叉路口和小半径的道路施工中,单车道滑膜摊铺机是非常理想的施工机械。

②双车道自动找平滑模式摊铺机采用了六步法调整,成形和抹平混凝土,从而获得无与伦比的质量和精度。

Part 2　Engineering and Construction Machineries

六步法连续摊铺：

（1）螺旋布料器分料；

（2）通过初步供料计量来调整进入振动器仓的混合料的流量；

（3）内部振动器振捣混合料；

（4）当水泥浆和骨料混合时，即在混合料到达熨平板之前，二次供料计量重新配比并且振捣混合料。

（5）两级振动熨平板（初级、次级；内部、外部）振捣混合料，形成最初的面板轮廓且初步抹平。

（6）配置磨边器的浮盘确定面板的最后形状。

滑膜摊铺机的工艺流程（图2-7-2）：螺旋布料器→虚方控制板→振动器→振捣棒→成型盘→拉杆插入器→调拱装置→浮动盘→磨边器→拖布等。

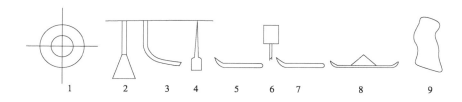

图 2-7-2　滑膜摊铺机工艺流程图

1-螺旋布料器；2-虚方控制板；3-振动器；4-振捣棒；5-成形盘；6-拉杆插入器；7-浮动盘（调拱）；8-磨边器；9-拖布

Key to the Exercises

Exercise One

1. 滑膜式摊铺机
2. 水泥混凝土摊铺机
3. 轨道式摊铺机
4. 成型盘
5. 浮动盘
6. 侧模版

Exercise Two

1. 振动器提浆以使混凝土密实和容易成型；而振捣棒则将大集料挤到面层下方以便成型盘形成均匀表面层。

2. 双车道自动找平滑模式摊铺机采用了六步法调整，成型和整平混凝土，从而获得无与伦比的质量和精度。

3. 在工作过程中，倾卸到摊铺机前端的水泥混凝土通过螺旋布料器沿着机器前端均匀的摊开。

4. 现代混凝土摊铺机大多数是滑膜式摊铺机，因为它的生产率高、结构紧凑。

第八课 压路机

①填铺在土方结构中的填土必须一层一层地压实,以获得所需要的密实度和稳定性。土壤可以自然密实,但是为了满足现代公路建设高速发展的需求,人们几乎完全摒弃了这种自然密实的方法,而是用机器来完成压实。压实机械也用于压实原状土,例如,压实挖方和水渠的底部及边坡,以减少由于渗透而造成的水流失。

压实机械在筑路工程中可用于压实(路基)稳定土。

压路机是通过滚轮对土壤表面施加作用力来完成压实的,压路机是应用最广泛、也是最简单的压实机械,它包括静力光轮压路机、羊角碾、轮胎压路机和振动压路机。

静力光轮压路机靠自重压实土壤,它可以压实路基、路面、广场和其他各类工程的地基。它可以分为:两轮两轴式;三轮两轴式和三轮三轴式。

轮胎压路机用来碾压、压实马克当(碎石)路面、沥青路面还有市政工程、土木工程和国家防御工程的地基。通过调整配重和轮胎气压可在压实沙质土和黏性土时获得较好的压实效果。在压实过程中,混合料的粒度将会改变且层与层间的内聚力也会更大。

如图 2-8-1 所示的四个前轮(转向轮)可以上、下移动;如图 2-8-2 所示的五个后轮(驱动轮)由左、右两个链轮带动。前、后轮胎交错安装、后轮压实前轮漏压的部分。

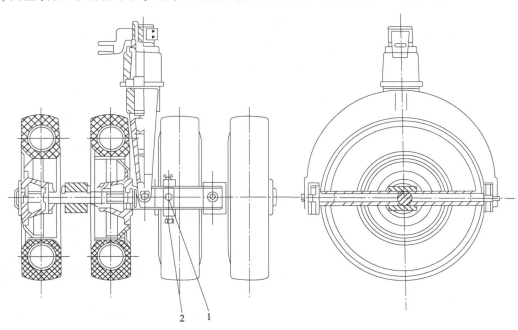

图 2-8-1 前轮
1-销子;2-螺栓

机器安装洒水系统,根据路面需求洒水。

配重(铁)块安装在车身下面来调整压路机的重量。

BW212 系列的单轮振动压路机是一种用于土方工程的自行式土壤压实机械。它靠振动器的往复运动使土壤颗粒移动实现压实。不同型号的压路机在压实能力和爬坡能力都有所不同。

振动压路机由柴油机驱动行走机构、振动机构以及恒速转向机构组成。

振动压路机的机架由带刮土器的前机架和后机架组成,前、后机架通过铰接相连。铰接可以用行驶锁锁死。

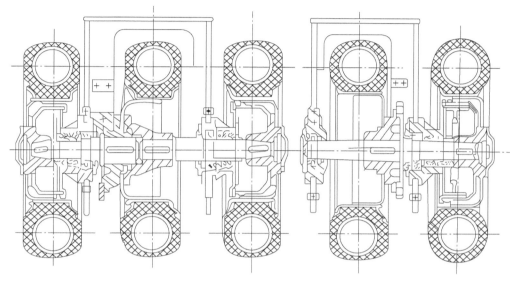

图 2-8-2　后轮

左侧安装振动马达和刮土器的振动轮柔性地悬挂在前车架上。后车架则用于支承驱动行走机构、振动机构以及转向机构的柴油机、液压油箱、燃油箱、驾驶室、仪表盘、蓄电池上的座椅以及驱动轮的行驶马达。后机架支撑在驱动桥上。

柴油机驱动行驶泵(容积变量泵),它给后驱动轮的驱动马达提供动力。驱动轮是胶轮,轮内加水和氯化钙(或者氯化镁)以增加重量,另外,还要额外充气。

驱动轮的行驶泵也用来驱动振动轮的行驶马达。

柴油机通过一个柔性联轴器来驱动振动油泵,振动油泵给振动轮左侧的振动马达提供动力,通过偏心轴的旋转产生振动。

通过控制仪表盘左侧的开关来起振,且只能在发动机全速状态下起振。

振动油泵是双向容积变量泵,通过泵的油液反向流动来改变振频和振幅,因为油液的流向可以改变振动马达的转向感应(信号)。振动马达的转向可以改变固定偏心块与移动偏心块的相互位置,从而改变偏心质量(两偏心块质量叠加或抵消)和振幅。高振幅时低振频;低振幅时高振频。

振频选择器用来预先设定振频(高、低),选择器中间位置无振动。

柴油机驱动的转向油泵提供油压实现动力转向。

行驶过程中,通过液压制动器实现制动。驻车制动器是通过人工控制作用在驱动轴上的盘式制动器实现制动效能。

BW 212 振动压路机布局如图 2-8-3 所示。

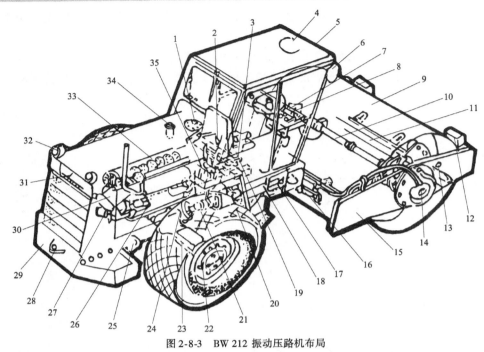

图 2-8-3　BW 212 振动压路机布局

1-液压油滤清管道；2-液压油箱视窗口；3-加热旋钮；4-换气装置；5-驾驶室；6-后视镜；7-振动马达；8-行驶控制杆；9-振动轮；10-偏心轴；11-刮土器；12-加热灯；13-前车架提升孔；14-振动轮振动马达；15-前车架；16-铰接锁；17-铰接装置；18-转向油缸；19-油门拉杆；20-驱动桥马达；21-驱动轴；22-行星齿轮；23-驱动桥；24-驻车制动器；25-后车架提升孔；26-转向油泵；27-振动油泵；28-牵引孔；29-后车架；30-干式空滤器；31-工具箱；32-工作灯；33-柴油机；34-燃油滤清器管道；35-定位楔块

Key to the Exercises

Exercise One

1. 作用在材料上的高压动力
2. 轮子内转动离心块产生的振动
3. 装满水的驱动轮
4. 固定偏心块
5. 移动偏心块
6. 静力光轮压路机

Exercise Two

1. 所有压路机都配有柴油机，用来驱动行走机构、振动机构以及恒速转向机构的液压泵。
2. 术语"频率"是指振动轮每秒钟的振动次数。
3. 振幅是指压实过程中振动轮"上、下"移动的量。
4. 通过稳定土法压路机也常用在筑路施工中。

第九课　沥青混合料拌和设备

沥青混合料广泛地用于筑路施工中，它是一种高等级路面材料。目前，沥青混合料的生产已经机械化,生产设备称为"沥青混合料拌和设备"。

沥青混合料拌和设备可以如下分类:

(1)根据施工时间的长短及施工地点的不同可分为:固定式、半固定式和移动式。

(2)根据工艺流程分为:间歇强制式和连续滚筒式。

1. 强制间歇式沥青拌和设备基本结构(原理)图

如图 2-9-1 所示,由皮带给料机从砂料斗输送的砂子和由电子振动给料机或皮带给料机从石料斗输送的石料被输送到连续式带式输送机上,再通过倾斜的带式(冷集料)输送机到冷集料烘干筒(除尘器)内。带式(冷矿料)输送机将冷集料送到烘干筒进料环,再将冷集料移动到一系列的(螺旋)叶片处。①随着烘干筒的转动,集料在燃烧器产生的热气中散落下来。并沿烘干筒长度方向移动到卸料口一端。推板(叶片)把热集料卸到卸料槽,再进入热集料提升机的料斗内。安装在底盘后部的多管旋风式、一级除尘器收集(烘干筒)的粉尘并通过两个螺旋输送机输送到烘干筒的卸尘槽内。热集料提升机把烘干筒的热集料运送到热集料筛分机中进行筛分。筛分后的集料按粒径大小分级并按级别分别储存在热储料仓的隔仓内。不能通过顶层筛网的大石块作为废料弃到拌和筛分装置一端的弃料槽内。

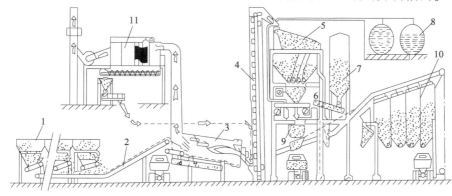

图 2-9-1 强制间歇式沥青拌和设备基本结构(原理)图

1-冷矿料储存及配料装置;2-冷料输送机;3-烘干筒;4-热矿料提升机;5-热料筛分装置;6-热料计量装置;7-粉料仓;8-沥青供给装置;9-拌和器;10-成品仓;11-除尘装置

粉料装在粉料仓内,循环供料的结合料(沥青)储存在沥青供料系统中。至此,拌和设备即可批量生产(混合料)。

在控制室里,操作员可以根据规范要求控制集料的流量。集料从储料仓卸到称量斗内并称量,称量后的再卸入一个完全封闭的提升机内,再送到搅拌器内。粉料通过搅拌器旁边的称量斗从粉料仓添加到搅拌器内,结合料(沥青)进入到搅拌器旁边的称量筒内称量。骨料进入到拌和器经过一段设定的干拌和时间,再加结合料继续进行一段湿拌和时间,直到这批混合料拌好卸到位于拌和器卸料门下方的运输机械中。根据需要,粉料可在结合剂添加之前或者之后添加。

2. 冷集料供料装置(冷集料给料机)

②冷集料供应装置是为装载式供料设计的,并由分离的砂料仓和石料仓组成,每个料仓都带有各自的用螺栓连在一起的支撑结构和出料机以形成一个整体。

砂料斗配置带式给料机,石料斗即可配置带式给料机也可以配置(电磁)振动给料机。带式给料机使用了无级变速驱动器。给料机把冷料输送到水平(安装)的槽式集料输送机上,然后再把冷料输送到倾斜的集料带式输送机上,倾斜的带式输送机把冷料供料机与烘干筒相连。

每个料斗的出口处都安装了可调的斗门,用来调节供料量并且可以清理(集料)堵塞。

冷集料给料机和集料输送机都独立安装了电机用来驱动。

3. 烘干/除尘器

烘干筒/除尘器使用了一个倾斜的集料输送机通过较宽的槽型皮带把集料从冷集料供料装置输送到烘干筒。使用的除尘器是多管旋风式、一级除尘器,并且安装了由电机驱动的叶片式风扇。通过安装在(烘干筒/除尘器)装置内的两条螺旋式输送机把汇集的尘土输送到烘干筒的卸尘槽内,也可以(通过风扇)吹到粉料仓内用来控制粉料(给搅拌器的)添加量。

③烘干筒由重型轧制钢板制成,并且沿着纵向交错安装有提升叶片。烘干筒安装的两条机械加工的钢制滚圈支撑在滚圈架上,在滚圈和滚圈架间预留间隙防止膨胀卡死。受料区从(倾斜的)集料输送机受料,再把集料提升到烘干筒内一系列的叶片处,集料在燃烧器产生的热气中散落下来。当热集料到达烘干筒另一端(排料端)时,推板把卸料端的热集料推到卸料斗内,卸料斗再把热集料送到热集料提升机的料斗中。

筒体通过一个电动机和完全封闭的齿轮装置由链条驱动,链条绕着筒体安装,主动链轮与电动机和齿轮装置相连。驱动机构使用柔起动机构使起动柔和,这样可以大大增加链条、从动链轮和主动链轮的使用寿命。

热量由安装在烘干筒出料端的自动配比的燃烧器产生。④自动配比燃烧器与按最新规范设计的燃烧室、鼓风机、油泵、管道、滤清器、阀和仪表做成一体,整个装置安装在一个可伸缩的托架上,以保证在维修时整个装置可以从烘干筒处移开。按标准配置带温度指示装置的遥控自动点火装置、人工火焰调节装置、风扇、火焰故障指示器。烘干筒的卸料槽使用了安装在控制室带遥控温度指示器的红外线温度计。

4. 拌和/筛分装置

拌和/筛分装置利用了热集料提升机从烘干筒的卸料槽接收集料进入到热提升机的接料斗中。料筒从接料斗中得到集料,再把它们提升到拌和筛分装置的顶部。

筛分装置顶部的弃料通过组装的钢制料槽卸到地面上。热料仓有3~5与隔仓,每个仓都配有溜槽把多余的集料输送到地面上。每个隔仓都在自行润滑的滚轴上安装了蝶形门,蝶形门靠两个汽缸控制,它可以使蝶形门以两级形式关闭,精确地控制称重量。称量斗悬挂在电子载荷传感器上,并且通过汽缸控制的全幅蝶形门把集料输送到全封闭的称量提升器中。

称量提升器使用了两排链条装置把集料从称量斗送到搅拌器。

搅拌器使用抗磨衬板全面加衬,包括铸钢拌和臂,(臂端)安装可更换的合金拌和桨叶。拌和臂安装在一对拌和轴上,轴安装在滚珠轴承内。搅拌器通过齿轮装置和电动机由两个重型链轮驱动。整个驱动装置安装了一个(防)超载装置,防止二次称量。

搅拌器使用了由汽缸控制的全幅旋转门,确保拌和器彻底清理和彻底卸料。拌和器还安装了低碳钢防尘盖,防止拌和时烟气和灰尘排放到大气中。

5. 粉料系统

粉料称量斗安装在搅拌器的上方,悬挂在电子称量传感器上。粉料称量斗通过气动的蝶形门卸料。

粉料储料仓的容量必须与拌和设备生产率以及各种规范相匹配。

通过螺旋输送机把粉料从料仓输送到称量斗。螺旋输送机通过电动机和柔性联轴器驱动。根据客户要求还可能提供一个或两个独立的料仓。

6. 沥青系统

结合料(沥青)称量筒安装在搅拌器的上方,悬挂在电子称量传感器上。

沥青箱可能电加热也可能油加热,并且按照生产量和沥青等级需求根据容量恒温控制。(称量的)沥青从沥青箱进入到称量筒内。称量筒恒温加热。沥青通过气动阀门从称量筒进入到沥青供料管,沥青供料管与支撑在搅拌器上的喷管连接。喷管把沥青喷到搅拌器中。

7. 搅拌器变形模块

此模块可以使强制间歇拌和或滚筒连续拌和设备成为一个独特的混合体,它结合了间歇拌和与连续搅拌设备的优点。

利用搅拌器变形模块即可以得到间歇搅拌设备所具有的高质量热混合料,也可以得连续搅拌设备所具有的高生产率。

在获得这两种设备的最佳性能的同时,还完全避免了每一种设备的先天缺陷。它可以将间歇搅拌设备转变成连续生产设备,就像一个连续搅拌设备一样。如果将其安装在连续拌设备中,就可以将搅拌过程与烘干过程分开,这样可以最大的速度生产更高质童的混合料。

Key to the Exercises

Exercise One

1. 沥青混合料拌和设备
2. 小型工地
3. 临时工地
4. 储料仓
5. 自行配比燃烧器
6. 电子载荷传感器

Exercise Two

1. 通过模块控制装置,每种拌和标准操作者都可以手工设定。
2. 粉料仓的容量应与设备生产量和规范需求相匹配。
3. 除了先前提到的间歇拌和设备,近些年来连续拌和设备也得到了广泛的应用。
4. 连续拌和设备的优点是简化生产过程和设备,因此减少了成本和环境污染。

第十课 冷铣刨机

铣刨机包括热铣刨机和冷铣刨机(图2-10-1)。安装加热装置的热铣刨机通常用在热再生路面的施工中。

冷铣刨机是沥青路面养护机械的机种之一,主要用于公路、城市道路、机场、货场等沥青混凝土路面的开挖翻修工程中。

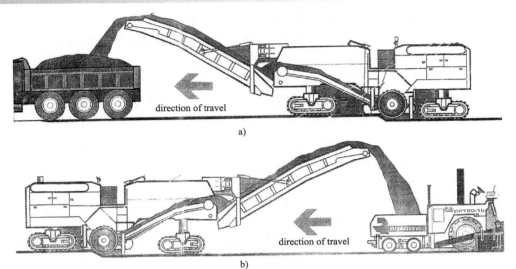

图 2-10-1　冷铣刨机
a)前卸(前部装载)下切式；b)后卸(后部装载)下切式

冷铣刨机还用来清除具有拥包、波浪、泛油、裂纹和车辙等病害的路面并且清除厚度可以满足重铺路面或重修路面的需求。它也可开挖路面坑槽及沟槽,还可以用于水泥混凝土路面的拉毛及面层错台的铣平。回收的沥青路面材料可以不需要进一步处理、而通过固定拌和设备的拌和再生使用。

目前,冷铣刨机种类繁多并用来满足道路处置的不同需求,可以对道路进行局部处置也可以进行彻底清除。

冷铣刨机包括动力装置、底盘、铣刨鼓总成、铣刨深度控制装置(自动找平系统)、液压系统、电器系统、装料系统和辅助系统。

铣刨机可以分为前卸(前部装载)上切式和后卸(后部装载)下切式。

1. 基本结构

W2100 铣刨机是一种履带行走、机械驱动铣刨鼓、具有两级前卸输送系统的铣刨机,可以调整装载高度并可向两侧转动。

2. 动力装置

铣刨机由柴油发动机驱动,发动机严格按照美国环境保护协会和欧盟标准排放。发动机安装全电子管理系统使发动机自动适应各种环境条件的变化,如大气压、温度和湿度。另外,发动机即使在极端负荷下,仍能提供最大转矩,确保施工不间断。

3. 铣刨鼓(图 2-10-2)

对于按上切铣刨工作的铣刨鼓来说,与点切刀具匹配的刀具架焊接在鼓体上。特殊(处理)的刀刃确保干净、利落的切削。辅助的抛物器确保铣刨材料有效地传送到主输送器(带)上。如果铣刨的材料遗留在路面上,刮板使它们积聚在履带之间。作为备选,铣刨鼓通常配备获得专利的、已经确定的快速更换刀具的系统。采用这种系统,刀具架的底座焊接在鼓体上,上部通过定位螺栓安装在基座上,这样可以迅速更换刀具。

铣刨鼓由柴油机通过机械离合器和作用在变速器的动力皮带机械驱动。由于采用了具有一定宽度的、12根的V形皮带,确保了最佳的动力传递,并且具有较长的使用寿命。驱动皮带的恒定张紧度由液压油缸来自动保持。

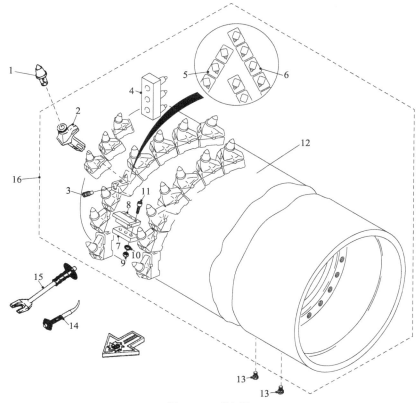

图2-10-2　铣刨鼓

1-铣削刀具;2-铣削上部刀座;3-定位螺栓;4-组合刀具;5、6-铣削刀具基座的布置;7、8-铣削刀具座;9-螺母;10-弹簧垫片;11-连接螺栓;12-铣刨鼓;13-螺栓;14、15-拌和刀具;16-罩壳

4. 刀具更换

①刮板由液压装置开启(可以转动100°),为更换铣刨鼓上的切削刀具提供了方便的通道。在两个后部履带行走装置之间为刀具架提供了足够的(安装)空间。

5. 装(卸)料

借助于由主输送器(带)和卸料输送器(带)组成的宽带输送系统,前卸式铣刨机可以更有效地将铣刨材料从铣削腔卸到前方的载货车上。

粒径控制梁可以在很大程度上防止沥青破碎成大厚片,同时也可以防止主传送带的早期磨损和撕裂。

②卸料输送器可以从很高的位置给载货车卸料,卸料的高度可调且可向两边转动,因此,总是能最佳的适应工地条件。

输送速度快、尺寸合理、具有足够的宽度、倾斜度较大的、加筋三角皮带保证将物料快速从铣刨仓中卸出。

卸料输送器(带)是封闭的,以免灰尘被风吹起而产生恼人的污染。输送系统的结构设计使皮带更容易更换。

6. 铣刨深度控制/自动找平系统

铣刨机装有电子自动找平系统来控制铣刨深度。它由比例阀控制,作用是相对于参考平面的高度变化可以得到迅速补偿并且不使机器产生超调。

③参考平面可以通过不同的方法进行扫描,例如,可以用侧板上的钢丝绳传感器扫描、旧路面上的超声波传感器扫描、对道路纵坡线使用角位移传感器扫描或用激光扫描。横坡传感器可以作为备选装置,连接件都要使用标准件。

7. 行驶系统与驱动

履带行驶装置通过液压调整高度油缸悬挂在底盘上,每条履带可以独立调整。铣刨深度通过两个前油缸(柱塞)设定,两条后履带完全作为浮动桥。油缸行程大使铣刨机离地间隙也大,因此即使在恶劣路况下也能安全操纵,比如在铣刨轨迹上安全倒车或安全地把机器开到低车架的拖车上。

铣刨机的行驶履带为重型履带。每一条履带由独立的液压马达驱动,马达由普通的液压变量泵供油。

履带行驶自动控制,所以,无须在铣刨和行走之间切换。行驶速度可以从零到最高无级变速。开关式液压分流器作为差速锁,保证在最困难条件下也能实现均衡牵引。

8. 转向系统

触摸操纵的全履带液压转向系统实现转向,可以在驾驶台的左侧或者右侧进行操作。它通过比例阀控制,通过操纵杆控制前、后履带独立转向。

9. 制动系统

自锁式液压变速器实现制动,铣刨机的前部还安装有两个自动的多片式驻车制动器。

10. 液压系统

行走装置、输送皮带、冷却风扇、喷水系统和调整油缸都由独立的液压系统驱动。液压泵由柴油机通过分动箱驱动。

11. 电器系统

电器系统电压为24V,安装有起动马达、三相交流发电机、两个12V蓄电池以及用于工作灯的插座。

12. 喷水系统

在铣刨过程中,液压驱动的喷水系统用来冷却点切铣刨刀具,因此,大大延长了刀具的使用寿命。水压和水量可以从操作台控制。喷嘴易拆卸方便清洗。

另外,主输送器(带)和卸料输送器(带)都是多点独立喷洒。铣刨鼓和传输系统的连续喷洒有效防止灰尘的漫延。

Key to the Exercises

Exercise One

1. 沥青路面
2. 冷铣刨机
3. 铣刨鼓
4. 加筋三角提升皮带

5. 卸料输送器
6. 差速锁

Exercise Two
1. 带加热装置的热铣刨机通常用在热再生路面的施工中。
2. 冷铣刨机用来清除病害路面达到所需要的深度。
3. 冷铣刨机可以分为前卸上切式和后卸下切式。
4. 卸料输送器应予以遮盖防止灰尘被风吹散造成烦人的污染。

第十一课　热再生机

热再生方法用于通过整形或添加黏结剂和新混合料修复出现裂纹、车辙、变形、老化和磨损的沥青黏结的路面面层。

热再生方法包括复拌再生(图 2-11-1)和重铺再生(图 2-11-2)。复拌再生是回收的旧料与新料重新拌和后再摊铺;重铺再生是把现有路面整形再重新摊铺一层新料。

重铺再生常用于破损严重路面的维修、翻新或旧路升级改造施工,而复拌再生常用来恢复道路的使用性能,两者可以完全利用现有的道路材料。

热再生机包括就地再生拌和设备和厂拌再生拌和设备。

如图 2-11-1,热再生机用于大面积沥青路面的修复工作。无极可变的施工宽度便于回收旧料、摊铺添加黏结剂和新料的混合料,可以连续进行(现场)加热、翻松、拌和、摊铺和碾压工。它包括加热器、路面再生机(重铺机、复拌机)、新料供料装置、搅拌机、混合料摊铺机、添料供料装置、熨平板及纵横坡控制装置。

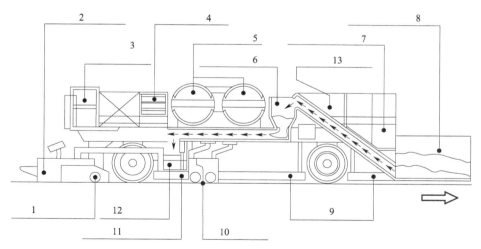

图 2-11-1　热再生机(复拌再生)
1-螺旋布料器;2-熨平板;3-控制台;4-柴油机;5-油箱;6-储料箱;7-沥青罐;8-接料斗;9-红外线加热器;10-可变宽度翻松器;11-红外线加热器;12-搅拌器;13-燃料箱

1. 加热

利用红外加热器对现有路面加热软化,混合料摊铺之前基层材料也被同时加热(热—热摊铺)。所用的热源是气态丙烷,蒸发气化后以气态形式燃烧。

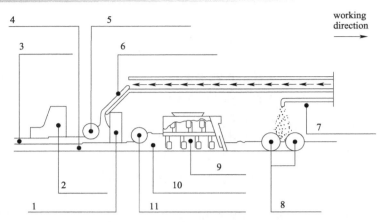

图 2-11-2　热再生机(重铺再生)

1-一次熨平板;2-二次熨平板;3-新料层;4-再生料层;5-二次螺旋布料器;6-新料供料器;7-沥青喷洒装置;8-翻松器;9-搅拌器;10-再生料;11-一次螺旋布料器

气罐:两个液化气罐,具有液面指示器。

汽化器:两个恒温控制的气动汽化器。

加热装置:全部为金属加热器,可以独立向外偏转以适应工作宽度的需要。

加热能量:通过设定工作压力,控制阀可以分配、控制每个燃烧器的输出热量。

2. 添料

实验室配制的添料由载货车批量装载到复拌机中,并连续地和一定比例的回收旧料(再生料)拌和。

接料斗:料斗具有标定容积,斗壁可实现液压倾翻。在卸料过程中,推滚推动自卸货车。

斜置输送器:具有耐磨的刮板和滚子链条的坚固的刮板输送器由液压驱动。输送器槽也应被加热,以免混合料冷却。

计量仓:为按钮式计量装置计量的添料提供储存容器,计量无级调整。

底盘输送器:具有耐磨的刮板和滚子链条的坚固的刮板输送器由液压驱动并无级可调。输送器槽也应被加热。添料或者进入拌和机(机门由液压开启),或者直接到熨平板前方。

自动供料控制:根据机器的前进速度,控制底盘输送器输送速度,即可实现预先设定的单位面积供料量(每平方米)。

3. 翻松装置

①翻松装置由两个滚筒式翻松轴和刀头组成,刀头将翻松的材料向内聚集,滚筒式翻松器和刀头将材料螺旋输送到搅拌器内。带齿的翻松器转动,翻松已被加热软化的路面,并在刀头的协助下螺旋的将松散材料输送到机器内部。通过液压系统无级调整施工宽度。

悬挂装置:翻松装置悬挂在位于机器两侧的轴上,并根据所需要的翻松深度由两个液压油缸定位。

翻松深度:设定的翻松深度由自动控制装置保持恒定。

4. 搅拌器

强制式搅拌器对再生材料和添料进行彻底的搅拌,直到形成均匀的混合料。

搅拌器:具有高强度衬片的水平双轴强制式(叶片)搅拌器由液压驱动,并且也被加热。

5. 熨平装置和找平控制

②搅拌好的混合料从搅拌器的出口卸出,并由无级调宽的熨平板精确地摊铺、整形。

振捣熨平板:液压驱动振捣棒和振动熨平板保证较高的预压实度。路拱可调。还可以选配一个辅助的熨平板,用来在事先已经喷洒沥青的旧沥青路面上进行精确摊铺。

熨平板坡度控制系统由两个纵坡传感器和一个横坡传感器组成。通过侧板或者张紧的钢丝绳感测参考平面。

液压驱动的清扫刷用来清扫熨平板前方(新旧)路面之间的接缝。

6. 沥青(再生添加剂)供给系统

沥青罐:因为安装控制温度的燃油加热器,沥青罐既可以装液态沥青也可以装块状沥青。

沥青泵:通过调整沥青泵速度可以无级地控制沥青添加量。

计量控制:根据机器的前进速度,自动控制装置控制沥青泵的速度,从而连续地控制预先选定的添加量(每平方米)。

喷洒装置:包括加热的喷管和喷嘴,沥青喷洒在双轴搅拌器前部,在搅拌器里充分搅拌。

Key to the Exercises

Exercise One
1. 斜置传送带
2. 铣刨深度
3. 沥青喷洒系统
4. 纵坡和横坡控制
5. 振动式熨平板
6. 红外线加热器

Exercise Two
1. 复拌再生是指把再生料与添料拌和后,重铺混合料。
2. 重铺再生法是指把现有路面整形,再铺新混合料。
3. 熨平板纵坡和横坡控制包括两个纵坡和一个横坡传感器。
4. 参考面或者用侧板或者用拉紧的钢丝绳感应。

>>> Part 3
专业英语的翻译方法与技巧

◎ 第一节　专业英语特点

◎ 第二节　翻译方法和技巧

专业英语与普通英语、文学英语相比有许多独特之处。专业英语与专业知识密切相关,除了包括数据、公式、符号、图表和程序外,在语言、语法、修辞、词汇、体裁等方面都有其独特之处,下面从语言、语法、词汇和结构上对专业英语的基本特点作一简要介绍。

第一节 专业英语特点

一、语言特点

(一)准确
译者必须把原作的全部科技内容准确、毫无遗漏地使用译文语言表达出来。

(二)简练规范
译者要用简洁易懂的语言表达原作的科技内容,避免不必要的修饰和重复。

(三)规范
所用的术语和词汇必须是本专业读者普遍应用的,句子结构符合汉语规范,语句精练、流畅。

二、语法特点

(一)非人称的语气和客观态度,常使用 It... 结构
专业英语所涉及的内容多为描述客观事实、现象和规律。这一特点决定了科技人员在撰写科技文献时要采用客观和准确的手法陈述被描述对象的特性和规律,研究方法和研究成果等,而不需要突出人物。因此,专业英语常使用非人称的语气作客观的叙述。

(二)较多使用被动语态
由于专业英语的客观性,决定了它非人称的表达方式。或者读者都知道动作的执行者是谁,或者不需要说明动作的执行者。因此,专业英语较多使用被动语态。

(三)大量使用不定式、分词和动名词
专业英语大量使用不定式、分词和动名词,目的是使句子简洁和精炼。

(四)较多使用祈使句和公式表达式
在理论分析和公式推导中常采用 Assume that...、Suppose that...、Let... 等表达方式。
如:Suppose that $P = 0$ at $x = y$
假设 $x = y$ 时 $P = 0$

(五)条件句较多
条件句多用于条件论述、理论分析和公式推导中,最常见的是 If... 条件句。

(六)长句较多
长句一般都是含有几个错综复杂关系的主从复句或并列复句,长句的重要特点就是修饰语较长,修饰语一般是短语或从句,它们是位于名词后面的定语短语或从句,或是位于动词后面状语短语或从句。具体来说,理解长句可以分为两个步骤:
(1) 判断句子是简单句、并列句还是主从句;
(2) 先找出句子的主要成分,即主语、谓语(系)和宾语(表语),再找定语、状语、宾补等。

(七)省略句较多

为了简洁,有时省略句子中的一些部分,如状语从句中的主语、谓语,定语从句中的关联词(which、that),但基本不省略形容词和副词。

三、词汇特点

每个专业都有一定数量的专业词汇或术语。一般分为三类:

(一)专业词汇

它的意义单纯就是专业含义,有时是根据需要造出来的。如,engine、chassis 等。

(二)半专业词汇

它大多数是各专业通用的,如,foundation(基础、基金、路基),frame(框架、车架、体系),load(装载、输入、充气、充电),operation(操作、手术、运行)

(三)非专业词汇

这类词汇是非专业词汇,但专业英语中常常使用,如,application、study、characterization。

(四)词缀和词根

据有关专家统计,现代科技专业英语中,有 50% 以上的词汇源于希腊语、拉丁语等外来语,而这些词构成的一个主要特征就是使用词缀(前缀、后缀)和词根,因此掌握一些词缀和词根,有助于扩大词汇量。

(五)缩写

现代科技专业英语中,一些专业词汇、政府机构、学术团体、科技期刊和文献都可以用缩写表示。如 ABS、CPU、EFI 等。

四、结构特点

一般在每一自然段落中,总有一个语句概括出该段落的重点。这个语句或在段落的句首或在段落的句尾。若干个自然段落会形成一个逻辑(结构)段落,用以从不同的角度来解说某一层面的核心内容。全篇则由若干个逻辑段落组成,从不同层面来阐述文章标题所表明的中心思想。

第二节 翻译方法和技巧

一、概论

(一)翻译的准则

翻译就是准确、完整地介绍原文的思想和内容,使读者对原文的思想和内容有正确的理解。翻译准则就是衡量译文质量的尺度,又是文章翻译所要遵循的原则。

对于翻译准则,一个比较传统的观点是:信、达、雅。"信"指准确、忠实原作;"达"指通达、顺畅;"雅"是指文字优美、高雅。由于专业英语本身注重表达技术问题的科学性、逻辑性正确性和严密性,所以,专业英语的翻译标准更侧重于信和达。

(二)翻译的过程
(1)深刻理解原文。
(2)弄清语法关系:即要按照原文的语言规范去认识每个句子中各种语法成分之间的关系。
(3)彻底辨明词义:
①多查阅几种不同的词典;
②把单词放到整句中去理解;
③从技术、专业内容去考虑。
(4)要以原文作整体,以段落作相对整体,以句子作表达单位,这样,才能保证意义上、语气上和逻辑上的有机联系。
(5)校对。
(6)修饰和润色。

二、词汇的翻译

(一)词类辨析
1. 分清词类
英语中的词可以根据词义、句法作用和形式特征分为:
(1)名词:表示人、物和抽象概念的名称的词。如:girl、city、war、notebook、honesty、family、Henry、London。
(2)形容词:用来描写或修饰名词(或代词)的一类词。如:active、soft、central、different、negative、famous。
(3)数词:表示数目多少或顺序先后的词。如:twenty-three、three hundred(and)forty one、first、one third、five and a half。
(4)代词:代替名词以及起名词作用的短语、分句和句子的词。如:you、us、your、yours、myself、each other、that、who、which、someone、all、both、either、neither、no、anything、little、few。
(5)动词:表示动作和状态的词。如:revolve、pave、move、look、smell、raise。
(6)副词:用以修饰动词、形容词、其他副词以及全句的词,表示时间、地点、程度、方式等概念。如:much、suddenly、rather、here、west、yesterday、down、too、so。
(7)冠词:置于名词之前、说明名词所表示的人或物的一种虚词。冠词只有三种:定冠词(the)、不定冠词(a,an)和零冠词。
(8)介词:一般置于名词之前,只表示其后的名词或相当于名词的词语与其他句子成分的关系。如:for、with、near、from、according to、out of、except、in。
(9)连词:连接单词、短语、从句、分句或句子的一种虚词。如:and、both and、because、if、as、if、or、that、whether。
(10)感叹词:用以表示喜怒哀乐等感情或情绪的词。如:oh,wow。
其中前六种可以在句子中独立担任成分,而后四种则不能。分清词类是正确理解词义的基础,也是准确表达原文的前提。

2. 根据后缀判断词类
英语中大量的词是同一词根加上一定的后缀派生出来的。这些词后缀是构成不同词类的标志。所以我们应该掌握各种词类常用的后缀,以便分清词类且能增加词汇量。

(1)构成名词的常见后缀:-er、-or、-ist、-ness、-(a)tion、-ment、-ion、-ing、-ure、-ty、-ity、-y、-th、-ance、-ence、-al 等;

(2)构成动词的常见后缀有:-ize、-en、-ify 等;

(3)构成形容词的常见后缀有:-ful、-less、-ish、-ive、-ous、-able、-ible、-ice、-ant、-al、-ent、-ary、-ory、-y 等;

(4)构成副词的常见后缀有:-ly、-ward、-wish 等;

(5)构成数词的常见后缀有:-teen、-ty、-th 等。

3. 根据词类理解词义

翻译时常常会遇到一词多义的英语单词,那么翻译时应根据上下文理解词义,准确判断词性。根据词在句中的词类来选择和确定词义。如:

1) round

(1) The clutch plate is round.

离合器是圆形的。(形容词)

(2) Round the corner slowly.

转弯慢行。(动词)

(3) The wheel turns round.

车轮旋转(副词)

(4) The wheel turns round the axle.

车轮绕轴旋转。(介词)

(5) Watchman makes his round every hour.

夜间值班员每小时巡逻一次。(名词)

2) Back

(1) The back wheels of the truck are turning.

货车的后轮正在旋转。(形容词)

(2) Put the tools back on the table.

把工具放回到工作台上。(副词)

根据上下文联系以及词在句中的搭配关系来选择和确定词义。如:

light music	轻音乐
light loss	轻微损失
light heart	轻松的心情
light steep	轻快的步伐
light outfit	轻巧的装备
light manners	轻浮的举止
light voice	轻柔的声音
light truck	轻型汽车

(二)词义的引申

(1)将词义作抽象化的引申:英语中,常常采用一个表示具体形象的词来表示一种属

性、一个事物、一个概念,翻译时可将其作抽象化引申。如:

The book is perhaps too high-powered for technician in general.

此书对一般的技术人员来说专业性太强。

high-powered 一般句中的意义是大功率。

(2)将词义作具体化引申:英语中也有用代表抽象概念或属性的词来表示一种具体事物,翻译时采用具体化的引申方法。如:

①The purpose of a driller is to cut holes.

钻床的功能是钻孔(不能翻译成切孔)。

②The car in front of me stalled and I missed the green.

我前面的那辆汽车熄火了,我错过了绿灯。

(三)词类的转换

在翻译过程中不能机械地按原有词类译出,需要词类转换,才能使翻译的文章通顺自然。

1. 转译成动词

英语中的名词、形容词、介词和副词,在翻译时均可转译成动词。

1)名词转译成动词

英语中大量由动词派生的名词和具有动作意义的名词以及其他名词,均可转译成汉语的动词。如:

Timely lubrication of all parts, proper tightening of all bolt and nuts and regular cleaning of truck are essential in ensuring its efficient performance.

及时地对各部件进行润滑,紧固所有的螺栓和螺母,并经常清洗汽车是保证汽车良好工作的必要条件。

2)介词转译成动词

英语中介词用得相当的多,某些含有动作意味的介词,翻译时往往译成动词。如:

Carrying capacity of the truck should be not over 5500kg in the running-in period.

在磨合期内,货车的载质量不应超过5500kg。

3)形容词转译成动词

英语中某些表示知觉、情感等心理状态的形容词与系动词一起构成复合谓语,往往可以译成汉语的动词,常见的这类形容词有:afraid、careful、anger、sure、certain、glad、sorry。如:

Are you familiar with the performance of this type of engine?

你熟悉这种类型的发动机的性能吗?

4)副词转译成动词

英语中,用作表语等成分的副词一般可译成汉语的动词。如:

She opened the window to let fresh air in.

她把窗户打开,让新鲜空气进来。

2. 转译成名词

英语中很多由名词派生的动词,以及名词转换的动词,在汉语中不易找到相应的动词,可将其转译成名词。如:

This chapter aims at discussing components of the asphalt paver.

本章的目的是讨论沥青摊铺机的各部件。

3. 转译成形容词

由形容词派生的名词或名词词组可译成形容词。如：

Vibration rollers have a variety of uses.

振动压路机有各种各样的用途。

（四）词的增译和省译

1. 词的增译

1）增加动词

根据意义上和上下文的需要，可以在名词前后增加动词。如：

(1) The opening of throttle blade is increased with the pressure of the foot paddle.

随着脚踏板压力的增加，节气门的开度也增加。

(2) High voltage is necessary for long transmission line, low voltage for safe use.

远距离送电需要高电压，安全用电则需要低电压。

2）增加形容词

根据专业知识和译文需求，有些句子可增加形容词。如：

(1) Oil passages in the head allow the oil to return to the sump by gravity.

汽缸盖上的油道使润滑油在重力的作用下回到油底壳。

(2) Some of gases in the air are fairly constant in amount, while others are not.

空气中有些气体的含量非常稳定，而有些不稳定。

3）增加副词

根据上下文，有些动词可增加副词进行修饰，使译文更贴切。如：

The manifold absolute pressure sensor responses to changes in the manifold pressure.

进气歧管绝对压力传感器及时、准确地感应歧管压力的变化。

4）增加名词

(1) 增加"现象"、"情况"、"过程"等词。如：

If the lubricating oil splashes to the driven plates, serious slippage will be happened.

如果润滑油溅到被动盘上，就会发生各种打滑现象。

(2) 增加概括性和解释性的词。如：

The units of "ampere", "ohm", and "volt" are named respectively after three scientists.

安培、欧姆和伏特这三个单位是分别根据三位科学家的名字命名的。

(3) 不及物动词后增加名词。有些不及物动词后实际隐含着宾语，翻译成汉语时往往把它表达出来。如：

All the major unit required to propel, direct, stop and run smoothly over varying road surface is known as chassis.

用来驱动车辆、使汽车转向、停车并且使汽车在道路不平的路面上平稳行驶的所有装置都称为底盘。

(4) 增加名词复数的词。如：

The chassis in general is equipped with a I-beam construction front axle.

通常,各种汽车底盘的前桥都是"工"字形梁。

5)增加承上启下的词

The clutch may be removed with or without removing the engine from the truck.

无论发动机是否从汽车上拆下,离合器都可以拆下来。

2. 词的省略

1)代词的省略

(1)省略作主语的人称代词

英语中,人称代词作主语往往多次重复出现,这种人称代词或泛指的人称代词翻译时可以省略。如:

①He was thin and haggard and he looked miserable.

他消瘦而憔悴,看上去一副可怜相。

②After the roller has operated a long time, we should inspect and maintain it in time.

压路机工作一段时间后,应及时检修和保养。

(2)省略物主代词

英语中的物主代词,翻译时往往可以省略。如:

The mass of one unit of volume of a material is called its density.

物体单位体积的质量叫作密度。

(3)省略非人称的或强调句中的"it"

①非人称"it"。如:

Before driving, check cooling water and various oils to see if it is necessary to renew or add fresh ones.

出车前,应检查冷却水和各种油料是否需要更换或添加。

②强调句中的"it"。如:

It is to increase the friction that the surfaces of automobile tyres are made with projections.

正是为了增加摩擦,才把汽车轮胎表面做成有许多突起的花纹。

2)冠词的省略

(1)省略不定冠词

不定冠词"a"和"an"在某些句子的翻译中,可以省略。如:

Fuel supply system plays an important role in the whole engine.

燃油供给系统在整个发动机中占有很重要的位置。

(2)省略定冠词

①用以专指某一个或某些人(事物),以别于同类的其他人(或物)、特别是该名词带有限定性定语或是同一文章再次提到时,"the"可以省略。如:

a. When the load of diesel engine is reduced, the speed is increased.

柴油机负荷减小时,转速会升高。

b. The movement of electrons is called electric current.

电子的运动称作电流。

②用于单数可数名词前面,泛指一类事物时,"the"可以省略。如:

The gearbox and clutch are made in an integral unit.
变速器和离合器制成一体。

③其后的名词是世界上独一无二的事物或江河、海洋、山脉、群岛等专用名词时,"the"可以省略。如：the Atlantic Ocean、the sun。

④用在重量、时间等名词前,表示一个单位时,"the"可不译或译为"每"。如：
This car travels twenty miles to the gallon.
这辆汽车每加仑汽油可行驶 20 英里。

（五）专业术语的译法

专业术语（specific terminology）指在各个学科领域中使用的专门术语。术语的特点是稳定、单一,如 hydrogen、oxygen 等。

1. 术语的构成

（1）派生术语：用前缀和后缀与基本词构成的新术语,如：

①表示数字和数量的前缀：semi-(半)mono-(一、单一)bi-(两)tri-(三),multi-(多)。

②表示学科或物件关系的常用前缀：electro-(电)、photo-(光)。

（2）复合术语：由两个或两个以上的单词组合构成的,如：
mainshaft(主轴)、dashboard(仪表盘)。

（3）缩写术语：这种术语是取几个词的第一个字母或词的一部分组成的新词,如：radar(radio detecting and ranging)雷达 laser(light amplification by stimulated emission of radiation)激光 U.F.O(Unidentified Flying Object)不明飞行物 GPS(Global Position System)全球定位系统 CPU(Central Processing Unit)中央处理器

（4）字母表示术语：这种术语是在基本术语前用英文字母的形状表示物体的几何形状,如：V-belt(三角皮带)、U-pipe(U形管)、I-beam(工字梁)、U-steel(槽钢)。

有一些字母不表示形状,而是表示一定的概念,如：
X-rays(X-射线)、Q-meter(Q表)、Q-band(Q-波段)、N-pole(N-极)。

（5）扩展术语：利用普通词汇的某一特殊意义而使该词成为专业词汇。如：
bearing 轴承(bear – 承受)、charge 充电(装载)。

2. 术语的翻译

1）意译

意译是根据原文术语运用汉语的构成要素,按照汉语的构词法创造出相应的新术语,意译是创造或翻译新术语的基本方法。如：semi-metal 半金属、dial-balance 刻度天平、teleprinter 电传打字机、photoflash 闪光灯。

2）音译

音译就是把原词的语音用标准汉语标出来,主要用来翻译以人名命名的科技度量单位以及人名、地点、某些货币单位。如：
joule 焦耳、hertz 赫兹、ohm 欧姆、pound 磅、volt 伏特(伏)、ampere 安培(安)。

3）音意兼译

它是两种方法的结合。如：
neon lamp 霓虹灯、Morse code、莫尔斯电码、beer 啤酒、card 卡片。

4) 直接使用原文

在专业文献资料中,牌号、型号、专业社团名称和表示特定意义(设备),的字母可不译,直接使用原文。如：

ABS system(Anti-lock Braking System)制动防抱死装置

EFI(Electronic Fuel Injection)电子燃油喷射系统

AC(Alternative Current)交流电

ADS(Association of Diesel Specialists)(美)柴油机专家协会

ISO(International Standard Organization)国际标准组织

三、不定式、分词和动名词

(一)不定式

不定式有两种形式:带 to 的不定式,其形式为"to + 动词原形",其中 to 仅是个符号,本身无实意;不带 to 的不定式,其形式就是动词原形。不定式在句中可作主语、宾语、补语、表语、定语和状语,起着多种语法功能。由于功能不同,其译法也不同。

1. 作主语

To remember to turn off the electricity is important.

记着关掉电源很重要。

2. 作宾语

During warm idle, igniting time is used to control idle speed.

在怠速暖车期间,用点火正时来控制转速。

有些带 to 的不定式结构前面加上疑问词后一起作句子的宾语,疑问词可以是 what、where、who(m)、when、how、whether、which 等。如:

This unit is used to check engine how to operate.

这个装置用来检测发动机的工作情况。

3. 作补语

有些动词的宾语之后还用动词不定式作宾补,此结构为"动词 + 宾语 + 不定式",在这种结构中,宾语是不定式的逻辑主语。如:

(1) Friction forces a motive body to stop.

摩擦迫使运动的物体停止运动。

(2) We consider him to be a good teacher.

我们认为他是一个好老师。

4. 作表语

The function of the steering system is to convert the rotary movement of the steering wheel in the driver's hands into angular turn of the front wheels, and to multiply the driver's effort.

转向系的功用是把驾驶员手中转向盘的转动转换成车辆前轮的转动并且放大驾驶员的作用力。

5. 作定语

不定式作定语时常位于所修饰的名词之后,翻译时应译在所修饰的名词之前。如:

Resistors require a reference voltage to work.

电阻器需要一个工作参考电压。

6. 作状语

不定式作状语可以修饰动词、形容词或副词,表示目的、结果、原因、条件等。如:

Modern digital computers operate fast enough to change injector pulse width in fractions of a second to maintain precise fuel metering.

为了精确的燃油控制,现代数字计算机运行的速度快到足以在零点几秒内改变喷油的脉冲宽度。

(二)分词

1. -ing 分词

-ing 分词由"动词原形 + ing"构成,称现在分词。-ing 不能单独构成谓语动词,但可以担任其他的成分。

(1)作主语。

Spraying precisely the same amount of fuel directly into the intake port of each cylinder eliminate the unequal fuel distribution so inherent when already mixed air and fuel are passed through an intake manifold.

精确地将等量的燃油直接喷入每个汽缸的进气口,消除了(单点喷射)混合气经过进气歧管所固有的燃油分配不均的现象。

(2)作表语。

The function of brake system is slowing down or stopping the car.

制动系的功用是减速或停车。

(3)作定语(常表示主动或现在)。

-ing 可单独作定语,也可由-ing 分词短语作定语。前者位于所修饰词的前面,后者位于后面。如:

①A transmission is a speed and power changing device.

变速器是速度和动力的变换装置。

②This is performed by the steering linkage consisting of pitman arm, a tie rod, a drag link, a steering arm and a left and right knuckle arms.

(转向)由包括转向垂臂、横拉杆、纵拉杆、转向臂和左右转向节臂组成的转向联动装置完成。

(4)作状语。

-ing 作状语表示时间、原因、条件、方式和伴随动作等意义。如:

①The intake valve closes and the piston moves up in the cylinder, compressing the air and fuel mixture.

进气门关闭且活塞在缸内上移,压缩空气燃油混合气。

②Starting effort depends on engine temperature, growing higher with decreasing temperature.

起动力取决于发动机的温度,温度越低,作用力越大。

2. -ed 分词

-ed 分词由"动词原形 + ed"构成,称过去分词。-ed 分词不能作谓语,在句中可以任其他成分。

（1）作定语（常表示被动或完成），如：

The chassis frame and units attached to it are supported on the wheels and tyre assemblies through front and rear suspension system.

车架和与车架相连接的装置通过前后悬架系统支撑在车轮和轮胎总成上。

单个的过去分词常作前置定语。如：

All the broken bearings have been replaced.

所有的坏轴承都已更换了。

（2）作状语。

过去分词作状语可以表示时间、条件、原因、伴随动作等。这种结构的逻辑主语就是主句的主语，前面往往加连词 when、if、as、while、though 等。如：

This field strength change repeated as each tooth passes the magnet, induces voltage in the wheel speed sensor coil and cable.

当每个齿通过磁场时，磁场强度的重复变化在车速传感器的线圈和导线中感应出电压。

（3）作宾语。如：

This will usually result in increased tire wear, reduced fuel mileage.

通常这将导致轮胎磨损加剧，里程数减小。

（三）动名词

动名词（动词原形＋ing）兼有动词和名词的特征，在句中主要起名词作用，作主语、表语、宾语和定语。

（1）作主语。如：

Lubricating is a means for reducing the friction between two moving parts.

润滑是减小运动件之间摩擦力的一种方法。

（2）作表语。如：

Reading is learning and applying is also learning.

读书是学习，使用也是学习。

（3）作定语。如：

The purpose of the cooling system is to keep the engine at its most efficient operating temperature at all speeds under all driving conditions.

冷却系的作用是在所有的驱动条件下和转速下保证发动机最有效的工作温度。

（4）作宾语。如：

The spring-mounted suspension softens shocks and ensures smooth running.

弹簧悬挂装置减小了（车辆）的振动，确保平稳行驶。

（5）动名词后面如果跟有宾语和状语等，便构成动名词短语，如动宾结构：

In laying a road bed bulldozers are employed to do all of works.

在铺筑路基时，推土机用来做各种铺筑工作.

四、句子的翻译

句子是语言的基本单位，翻译时基本上也是以句子为单位的。因此，对一个句子里包含的词语或从句,也必须按特定的句子结构和语言环境进行分析和判断,并且,要遵从汉语的语法修辞习惯,这样,才能做到准确地表达原义。

(一)简单句

简单句有"主语+谓语"、"主语+连系动词+表语"、"主语+谓语+宾语"、"主语+谓语+间接宾语+直接宾语"、"主语+谓语+宾语+宾语补语"等五种基本结构。

(1)"主语+谓语"句型可简称主谓结构,谓语是不及物动词。如:

The temperatures change.

温度是变化的。

(2)"主语+连系动词+表语"句型可简称主系表结构。如:

The pressure and temperature at the end of compression become very high.

压缩终了(汽缸)内的压力和温度都非常高。

注:连系动词后的表语可以是名词、形容词、分词、介词短语等。

(3)"主语+谓语+宾语"句型可简称为主谓宾结构,谓语一般是及物动词,宾语一般是直接宾语。如:

The brake retards the motion of the car.

制动器阻止汽车的运动。

(4)"主语+谓语+间接宾语+直接宾语"句型可简称为主谓宾宾结构,谓语一般是及物动词,宾语一是间接宾语,一是直接宾语。如:

The cooling system provides the engine cooled water.

(5)"主语+谓语+宾语+宾语补语"句型可简称为主、谓、宾、宾补结构,谓语一般是及物动词,宾语与宾补构成复合宾语。如:

We call that I-beam girder.

我们把那根"工"字形的梁叫(车架)大梁。

注:"主语+谓语+宾语+宾语补语"和"主语+谓语+间接宾语+直接宾语"句型中,前者宾语和宾补指同一事物;而后者间接宾语和直接宾语指两种事物。

(二)并列句

由两个或两个以上的简单句并列而成的句子叫并列句。常见的形式是:简单句+并列连词+简单句,各个简单句通常用并列连词连接起来。

(1)并列连词表示连接(and)、转折(but)、选择(or)、因果(so)等。如:

①Sometimes chassis is known as the carriage unit and includes the instrument panel.

有时底盘也称为支撑装置并且包括仪表盘。

②Internal friction cannot be eliminated, but it can be reduced by use of lubricants.

(发动机)的摩擦不可能消除,但通过使用润滑剂可以减小。

(2)有时不用并列连词,而用逗号、分号、冒号把分句分开。如:

The water pumps consist of rotating fan, seldom are of the displacement type of gears.

水泵是叶片式泵,基本不用齿轮式容积排量泵。

(三)特殊句

科技英语中,经常出现被动句、否定句、强调句等句型。这些句型都有其自身特点,往往和汉语句型有些不同,在翻译时容易造成错误,因而要特别注意。

1. 被动句

与汉语相比,英语中被动句使用的范围相对要广。凡是出现以下情况,英语常用被动语态。

①不必说出行为的执行者;

②无从说出行为的执行者;

③不便说出行为的执行者。

1)译成汉语主动句

Friction can be reduced and the life of the machine prolonged by the lubricant.

润滑油可以减少摩擦,增加机器寿命。

2)译成汉语被动句

原句中的主语仍译成主语,而被动意义用"受到"、"使"、"由"等词来表达。

The operational delays are caused partly by the internal frictions within the traffic streams themselves.

运行延误的部分原因是车流内部相互间的阻力造成的。

2. 否定句

英语中的否定句多种多样,与我们熟知的一般否定形式不同的是,英语中有一些特殊的否定句,其否定形式与否定概念不是一致的,它们所表达的含义、逻辑都与我们从字面上理解的有很大的差别。总之,英语中的否定问题是一种常见的且又复杂的问题,值得特别重视。

1)否定成分的转译

否定成分的转译是指意义上的一般否定转为其他否定,反之亦然。常见句型有:

(1) not...so...as...结构

在谓语否定的句子中,如果带有so...as连接的比较状语从句,或as连接的方式状语从句,就应该译成"不像……那样……",而不能直译成"像……那样不……"。

①The sunrays do not warm the water so much as they do the land.

太阳光线使水增温不如它使陆地增温那么多。

②Friction is not always a bad thing as you might think.

摩擦并不是你想的那样总是坏事。

(2) not...think/believe...结构

表示对某一问题持否定见解时,此句型要译成"认为……不……","觉得……不是……"。

Ordinarily we do not think air as having weight.

一般情况下,我们认为空气没有重量。

(3) not...because...结构

这种结构的否定句可以表示两种不同的否定含义,既可以否定谓语也可以否定原因状语because,翻译时要根据上下文意思判断。

The version is not placed first because it is simple.

这个方案因为简单不能放在句首。或者:

这个方案不因为简单而放在句首。

2）部分否定的译法

英语句子中常用 not 与 all、both、every、many、much、often、always 搭配使用表示部分否定,一般译成"不是都"、"不总是"、"不全是"。

（1）All corrosion is not caused by oxidation.

腐蚀不都是由氧化反应引起的。

（2）Both of the dams are not gravity dams.

这两座大坝不都是重力坝。

3）否定语气的改变

英语中的否定句并非一概译成汉语的否定句,有些否定句表达的是肯定的意思,常见的是 nothing but（只有,只能）句型;也有些否定句在特定的语义环境下表达肯定的意思。

（1）Early computer did nothing but compute：adding, subtracting, multiplying and dividing.

（2）早期计算机只能做加、减、乘、除运算。

4）意义否定的译法

有些句子没有出现否定词,但句子中出现了表示否定的词和词组,翻译时要译成汉语的否定句。如：

Such glass would bend like metal when dropped rather than shatter into bits.

这种玻璃掉在地上时会像金属一样弯曲而不碎裂。

在科技英语中,常见的含有否定意义的词组有：

Beyond	无,没有	above	不
but for	如果没有	too...to	太……不
free from	没有	too...for	太……不
instead of	而不是	fail to	不能
in the absence of	没有……时	far from	远非
short of	缺少	safe from	免于
but that	要不是		

5）双重否定的译法

（1）针对同一事物的否定

语法否定（not）和语义否定（没有 not,但有否定意义的词或词组）都针对同一事物而言,即"否定的否定"。

With a careful study of all the data made available to this engineer, there could be nothing unexpected about problem.

通过这位工程师对所有资料进行仔细研究,这个问题所涉及的一切都在意料之中了。

（2）针对两种不同事物的否定

两个否定分别针对两种不同的事物,不是"否定的否定",只是一句话里有两个否定含义的词汇而已。如：

There is not steel not containing carbon.

没有不含碳的钢。

3. 强调句

强调句型"It is(was) + 强调部分 + that(which,who)..."几乎可以用于强调任何一个陈述句的主语、宾语或状语。需要注意的是此强调句型与带形式主语的主语从句很相似,但不同的是去掉几个英文单词后,强调句剩下的单词仍能组成一个完整的句子。

1)强调部分是主语或状语时,一般不改变原文句子结构

It is drawbacks that need to be eliminated and have let to the search for new process.

正因为有这些缺点需要消除,才导致了对新处理方法的研究。

2)当被强调部分是宾语时,必须改变原原文句子的结构

翻译时,把所强调的宾语译成表语,而把原来的主语、谓语合译成汉语的"的"字结构作主语,即译成"的是"、"的正是"、"的就是"。如:

It is the losses caused by the friction that we must try to overcome by various means.

我们必须想办法来克服的正是由摩擦引起的损失。

(四)从句

1. 主语从句

用作主语的从句叫作主语从句。

(1)引导主语从句的关联词有疑问代词,疑问副词,缩合连接代词,缩合连接副词等。如:

①Which of the three forms a substance will take depends on the kind of substance, and on its volume, temperature, and pressure.

物质取三态中的任一种取决于物质的种类、体积、温度和压力。

②Where the failure of this roller is is not clear to anyone.

这台压路机的故障出在哪里谁也不知道。

③How well a diesel engine can operate with different types of fuel is dependent upon engine operation conditions, as well as fuel characteristics.

发动机使用哪种燃油运行得更好取决于发动机的运行条件和燃油特性。

(2)以 it 作形式主语引出的主语从句,it 可以译出也可以不译出。如:

①It is essential that the sub-base should be compacted to a uniform density, since the density of a soil is closely related to its bearing capacity.

路基压实达到均匀的密实度是非常必要的,因为土壤的密实度与承载能力密切相关。

②It will be noted that both inlet and exhaust valves are closed during in compression stroke.

值得注意在压缩过程中进、排气门都关闭。

③It is at the middle point that engine output efficiency is its maximum.

(曲线)的中点是发动机最大输出效率值。

2. 表语从句

用做表语的从句叫表语从句。

(1)引导表语从句的关联词有疑问代词、疑问副词、缩合连接代词、从属连词等。如:

①Pure copper wires have very low resistance, which is why they are so much used in electric circuits.

纯铜线的电阻很低,这就是为什么在电路中大量使用它们的原因。

②In an automobile the torque or crankshaft twist developed by the engine is what turns the drive wheels.

在汽车上,正是发动机产生的转矩驱动车轮。

(2) 从属连词 that 有时也可引导表语从句。如:

Another important point about EGO sensors is that they measure oxygen.

氧传感器另一个重要的特点是可以计量氧含量。

(3) 从属连词 whether 有时也可引导表语从句。如:

The question is whether the engine is operating well.

问题是发动机是否运转良好。

3. 宾语从句

在句中用做宾语的从句叫作宾语从句。

(1) 引导宾语从句的关联词有从属连词、疑问代词、疑问副词、缩合连接代词和缩合连接副词等。如:

①The experience has shown that the system is best suited for engines with small cross-section manifold runners.

实践证明该系统最适用于进气歧管为小断面结构的发动机。

②The computer monitors the voltage drop and by using a look-up chart of stored pressure value, determines exactly what the pressure is to which the diaphragm is responding.

计算机监测电压降,并通过查阅储存的压力图表,准确地确定硅片所反映的压力值。

③Multipoint injection is often referred to as port injection and means that fuel is introduced into engine from more than one location.

多点喷射指气道喷射,表示燃油从多点喷到汽缸。

(2) 宾语从句可用做介词的宾语。如:

①As the current crosses the resistor/diaphragm, the amount of voltage drop that occurs depends on how much the diaphragm is flexed.

当电流通过电阻/膜片时,电压下降的量取决于膜片弯曲的程度。

②The sub-base will vary in depth according to the nature of the sub-grade, and also according to what thickness of concrete is to be laid above it.

路基的厚度是根据地基的自然属性和上面所铺的沥青混凝土的厚度来确定。

③The choice of plant used will depend on how deep a cut is required, and also on how accessible the cut is.

设备的选择取决于挖方的深度还取决于取土的难易程度。

(3) It 作形式宾语。如:

①The growth of traffic made it necessary that we solve the problems scientifically.

交通量的快速增长需要我们科学地解决这一问题。

②This makes it possible that catalytic converter works effectively to reduce exhaust.

(计算机控制)使催化转换器有效的工作减小污染排放成为可能。

(4) 现在分词后可跟宾语从句。如：

He has just gone away saying that he will return in an hour.

他刚走，说一小时后回来。

4. 定语从句

在句中作定语的从句叫定语从句。定语从句通常位于它所修饰的名词(代词)之后，这种名词(代词)叫先行词。引导定语从句的关联词为关系代词和关系副词。关系代词在句中可作主语、宾语、定语等。关系副词在句中只用做状语。

(1) 关系代词有 who、whom、whose、that、which 等。who 是主格，在句中作主语；whom 是宾格，在句中作宾语；whose 是所有格，在句中作定语。如：

①This men who is working is a engineer.

正在工作的那个人是工程师。

②I know the teacher whom you mean.

我认识你指的那个老师。

③That is the driver whose pick-up truck failed.

这就是轻型货车出故障的那个驾驶员。

(2) that 在句中即可作主语，也可作宾语。即可指人，也可指物，但多指物。

①The amount of fuel delivered by the injector depends on the amount of time that the nozzle is open.

喷油器的喷油量取决于喷嘴打开时间的长短。

②The ideal spring for automobile suspension would be one that would absorb road shock rapidly and then return to its position slowly.

汽车悬架系统理想的弹性元件应该是迅速吸收振动，然后缓慢复位。

(3) which 在句中即可作主语，也可作宾语。一般指物。(先行词指物时 that 和 which 往往可以互换)

①The shaft of the gasoline engine carries a pinion which can be meshed with teeth on the diesel engine flywheel.

汽油机的输出轴上安装了一个小齿轮，它与柴油发动机的飞轮环齿相啮合。

②Kinetic energy is the energy of motion which is converted into heat given up to air flowing over the braking system.

动能是指物体运动的能量，它可以通过流经制动系统的空气转化成热向外散出。

(4) 用作关联词的关联副词有 when、where、why 等。when 在从句中用作时间状语，其先行词须是表示时间的名词；where 在从句中作地点状语，其先行词须是表示地点的名词；why 在从句中作原因状语，其先行词须是表示原因的名词。如：

①The time when the current takes to heat the resistance winding is the timing factor of the switch and determines the time that the circuit is closed to operate the cold start injector.

电流什么时间加热电阻线圈是开关的定时参数，并且决定了使冷起动喷嘴工作电路的闭合时间。

②The secondary venturi creates an air stream, which holds the fuel away from the barrel walls where would slow down and condense.

第二喉管产生气流使得燃油离开可能引起燃油减速和冷凝的化油器内腔壁。

③This is the reason why multipoint injection is provided.

这就是为什么采用多点喷射的原因。

（5）非限定性定语从句。这种定语从句对先行词不起限定作用,只是对整个句子加以补充说明或描写、叙述、解释,在书面语中用逗号分开。（限定性定语从句与先行词关系密切,对它起限定作用,不可缺少）如：

①The drive is accomplished by the friction between the three members, which depends upon the material in contact and the pressure forcing them together.

传动靠三个部件之间的摩擦力来完成,摩擦力取决于部件之间的接触材料和作用在它们上面的压力。

②The screed comprises two main sections to form the basic screed and extensions, which are built out to the required width.

熨平板包括两部分形成基本熨平部分和延伸部分,延伸部分用来保证摊铺所需要的宽度。

（6）关系代词在定语从句中作介词宾语,介词即可置于从句之首,也可置于从句之尾。

①The minimum speed with which the crankshaft of an engine should be rotated to ensure reliable starting of the engine is referred to as the cranking speed.

保证发动机可靠起动曲轴的最小转速称为发动机的起动速度。

②This variable percentage of the complete cycle during which the solenoid is on is the solenoid'duty cycle.

电磁线圈通电时间占整个循环的百分比叫电磁线圈的工作循环。

5. 状语从句

用作状语的从句状语从句。引导状语从句的关联词是某些从属连词。状语从句根据其用途可分为时间状语从句、地点状语从句、原因状语从句、结果状语从句、程度状语从句、目的状语从句、条件状语从句、让步状语从句和方式状语从句九种。

（1）时间状语从句:关联词有 as、before、once、since、until、when、whenever、while、as long as、as soon as 等。如：

①Spraying precisely the same amount of fuel directly into the intake port of each cylinder eliminates the unequal fuel distribution so inherent when already mixed air and fuel are passed through an intake manifold.

直接对每个汽缸的进气口精确地喷入等量的燃油,消除了已经混合好的混合气通过进气歧管时的本质上不均匀分配现象。

②The brake switch opens the supply voltage to the transmission solenoid whenever the brake is applied.

无论什么时间施加制动,制动器开关都会打开向变速器电磁线圈提供电压。

③The driver had to insert a crank into the front of the engine and turn it by hand until the engine started operating.

驾驶员向发动机的前端插入一曲柄并用手摇动,直到发动机正常运行为止。

④As soon as the solenoid valve moves to the reduce position, brake fluid from the wheel circuit flows into the accumulator.

电磁阀一向减小的位置移动时,轮缸制动液就会流到储液器中。

(2) 地点状语从句:关联词有 where、wherever、anywhere、everywhere 等。如:

The hot water from the water jacket is then through the top hose into a radiator where it is cooled by the air outside before it is pumped back through the bottom hose into water jacket again.

水套中的热水通过上水管进入到散热器中,在它由下水管泵回到水套之前由外部的空气对它进行冷却。

(3) 原因状语从句:关联词有 because、as、since 等。如:

①Since a internal combustion engine develops little power or torque at low rpm, it must gain speed before it will move the vehicle.

因为内燃机在低速时产生的功率或转矩都非常小,因此在车辆行驶之前必须达到一定的转速。

②Specialization is a desired goal for society because it increases efficiency and directly improves our standard of living.

专业化是社会需求的一个目标,因为它增加了社会效益并改善我们的生活水准。

(4) 结果状语从句:关联词有 so(that)、such that、that、with the result that 等。如:

Rubble hoses are liable to perish and crack, so they must be checked regularly and replaced if they are found to be damaged.

橡胶软管极易损坏且容易出现裂纹,所以要定期检查,如果发现损坏要及时更换。

(5) 程度状语从句:关联词有 so(that)、such that、as(so)far as、to the degree that、in so far as 等。如:

Finally, the engine reaches a degree that it can handle the vehicle and road load all by itself.

最后,发动机可以达到自行控制车辆和道路载荷的程度。

(6) 目的状语从句:关联词有 so、so that、in order that 等。如:

The driven member is made as light as possible so that it will not continue to rotate for any length of time after the clutch has been thrown out.

制作的被动件尽可能的轻,以便在离合器分离后不会旋转的时间过长。

(7) 条件状语从句:关联词有 if、unless、providing(that)、provided(that)、in the event (that)、given(that)、in case that、as(so)long as。如:

①If the upper chamber is connected to the intake manifold, the sensor functions as a manifold absolute pressure sensor.

如果上腔与进气歧管相连,此时传感器可以作为进气歧管绝对压力传感器。

②Given the person must perform some of these activities to survive, he/she needs transportation to accomplish his/her activities.

人们知道人类必须从事某些活动得以生存,活动必须依赖交通得以完成。

（8）让步状语从句：关联词有 though、although、if、even if、for all（that）、in spite of the fact that 等。如：

①The friction is the main factor governing even if sufficient effort is available to lock the wheels.

即使使车轮抱死的作用力足够大，但摩擦力仍是一个主要的控制因素。

②The pressure applied to the circuit is steady, even though the master cylinder pressure is higher as the brakes are applied.

尽管作用在主缸上制动力很高，但作用在轮缸上的作用力是稳定的。

（9）方式状语从句：关联词有 as、as if、as though、the way、how 等。如：

The engine injects fuel timely as if there is no failure at all.

发动机仍定时喷油好像没有任何故障一样。

6. 同位语从句

用作同位语的从句叫同位语从句。其形式与定语从句相似，两者之前都有先行词，但与先行词的关系不同：同位语从句与先行词同位或等同，定语从句则与先行词是修饰关系。

（1）同位语从句的先行词多为 fact、news、idea、though、question、reply、report、remark 等，关联词多用连词 that。如：

The actual significance of this technical innovation at that time is demonstrated by the fact that today, roughly 3 decades later, nearly all modern concrete pumps function according to the same principle.

那时这项技术革新的实际意义在于这样一个事实：即大约30年后的今天，几乎所有现代混凝土泵都按同样的原理工作。

（2）疑问代词 who、which、what 和疑问副词 where、when、why、how 也可引导同位语从句。如：

①This is the question where the oil leaks from.

问题是机油从哪里泄漏。

②This is the question when spark plugs produce igniting sparks.

问题是火花塞什么时间产生火花。

（3）同位语从句也可由 whether 引导。如：

These feedback signals tell the computer whether the output is too little, too much, or just right.

反馈信号通知计算机输出是否太小，太大或正好。

（五）复合句

由一个主句和一个或一个以上的从句构成。从句不能单独成句，但它有自己的主语和谓语。从句必须由关联词引导。引导从句的关联词有七类：

（1）从属连词：whether、when、if、because、although。如：

The lubrication system is classified as "pressure" and "splash" although various combination of these two systems are used.

润滑系统分为压力润滑和飞溅润滑，不过，这两种润滑是以多种结合的方式来使用的。

（2）疑问代词：who、whom、whose、which、what。如：

This system uses an intake manifold similar to what would be used with a carbureted engine.

系统使用的进气歧管与化油器发动机所使用的进气歧管类似。

（3）疑问副词：when、where、why、how。如：

The hydraulic system must know when the gearbox changes from one gear to another.

液压系统必须知道变速器什么时候换挡。

（4）关系代词：who、whom、whose、which、that。如：

Kinetic energy is the energy of motion which is converted into heat given up to air.

动能是指（物体）运动的能量，它可以转化成热释放到空气中。

（5）关系副词：when、where、why。如：

The engine reaches a point where it can handle the vehicle and road load all by itself.

发动机达到可以自行处理车辆和道路载荷的值。

（6）缩合连接代词：what、whatever、who、whoever、whichever。如：

Whichever the part fails, the diagnosis system can find it.

无论哪一个部件出现故障，诊断系统都可以诊断出来。

（7）缩合连接副词：whenever、wherever、however。如：

Whenever the part fails, the diagnosis system can find it.

无论什么时候部件出现故障，诊断系统都可以诊断出来。

附录 工程机械常用英文缩写

缩写	英文含义	中文含义
AAT	Ambient Air Temperature	环境温度
ABS(ALB)	Anti-lock Brake System	制动防抱系统
AC	Alternating Current	交流电
A/C	Air Conditioner	空调
Acc	Acceleration	加速
ACCEL	Accelerator	加速踏板
ACCRY	Accessory	附件
ACL	Air Cleaner	空气滤清器
ACT	Air Charge Temperature	进气温度
ACSD	Auxiliary Cold Start Device	冷起动辅助装置
ADC	Analog-Digital Converter	模拟–数字转换器
ADS	Association of Diesel Specialists	(美)柴油机专家协会
A/F	Air Fuel Ratio	空气燃料比
AFC	Air Flow Control	空气流量控制
AFS	Air Flow Sensor	空气流量传感器
AI	Air Injection	空气喷射
AIV	Air Injection Valve	空气喷射阀
ALCL	Assembly Line Communication Link	总装线测试插座
ALDL	Assembly Line Diagnostic Link	总装线诊断插座
ALT	Alternator	交流发电机
A/M	Automatic/Manual	自动/手动
ANT	Antenna	天线
APS	Absolute Pressure Sensor	绝对压力传感器
AP	Accelerator Pedal	加速踏板
APS	Absolute Pressure Sensor	绝对压力传感器
ASM	Assembly	总成
ASSY	Assembly	总成
A/T	Automatic Transmission(Transaxle)	自动变速器
ATDC	After Top Dead Center	上止点后

ATF	Automatic Transmission Fluid	自动变速器油液
AWD	All Wheel Drive	全轮驱动
B+	Battery Positive Voltage	蓄电池正极
BARO	Baromatric Pressure	大气压
BARO sensor	Baromatric Pressure Sensor	大气压力传感器
BP	Baromatric Pressure Sensor	大气压力传感器
BAT	Battery	蓄电池
BDC	Bottom Dead Center	下止点
BTDC	Before Top Dead Center	上止点前
CACS	Charge Air Cooler System	增压空气冷却系统
CCC	Converter Clutch Control	转换离合器控制
CD	Coefficient of Drag	(空气)阻力系数
CES	Clutch Engage Switch	离合器接合开关
CG	Center of Gravity	重心
CGND	Case Ground	外壳接地(搭铁)
CHG	Charge	充电
CID	Cylinder Identification Sensor	判缸传感器
CKP(CP)	Crankshaft Position	曲轴位置
CKP sensor	Crankshaft Position Sensor	曲轴位置传感器
CL	Clutch	离合器
CMP	Camshaft Position	凸轮轴位置
CMP sensor	Camshaft Position Sensor	凸轮轴位置传感器
CMFI	Central Multipoint Fuel Injection	中央多点燃油喷射
CMP	Camshaft Position	凸轮轴位置
CN	Cetane Number	十六烷值
CPC	Clutch Pressure Control	离合器压力控制
CPP	Clutch Pedal Position	离合器踏板位置
CPS	Camshaft Position Sensor	凸轮轴位置传感器
CPS	Crankshaft Position Sensor	曲轴位置传感器
CPU	Central Processing Unit	中央处理器
CTS	Engine Coolant Temperature Sensor	发动机水温传感器
CVJ	Constant Velocity Joint	等速万向节
CVT	Continuously Variable Transmission	无级变速器
CYL	Cylinder	气缸
CYP	Cylinder Position	气缸位置
DC	Direct Current	直流电
DCV	DC Voltage Source	直流电源
DCV	Deceleration Control Valve	减速控制阀

缩写	英文	中文
DCV	Double Check Valve	双单向止回阀
DID	Direct Injection-Diesel	柴油直接喷射
DIFF	Differential	差速器
DO	Diesel Oil	柴油
DOHC	Double Overhead Camshaft	顶置双凸轮轴
DS	Detonation Sensor	爆震传感器
DTC	Diagnostic Trouble Code	诊断故障码
DV	Delay Valve	延迟阀
EATX	Electronic Automatic Transmission/Transaxle	电控自动变速器
EC	Engine Control	发动机控制
ECA	Electronic Control Assembly	电子控制总成
ECD	Electronic Control Diesel	电子控制柴油机
ECI-MULTI	Electronically Controlled Injection Multipoint	电控多点燃油喷射
ECL	Engine Coolant Level	发动机冷却液面
ECM	Electronic Control Module	电子控制模块
ECM	Engine Control Module	发动机控制模块
ECT	Engine Coolant Temperature	发动机冷却水温度
ECU	Electronic Control Unit	电子控制单元
EEC	Electronic Engine Control	电子发动机控制
EFI	Electronic Fuel Injection	电控燃油喷射
EGR	Exhaust Gas Recirculation	废气再循环
EGOS	Exhaust Gas Oxygen Sensor	氧传感器
ELB	Electronically Controlled Braking System	电控制动系
EM machines	Earthmoving Machines	工程机械土方机械
ENG	Engine	发动机
EPS	Electronic Power Steering	电子动力转向
EX	Exhaust	排气
F	full	满的,全的
FBC	Feed Back Control	反馈控制
FC	Fan Control	风扇控制
FDR	Final Drive Ratio	主传动比
FE	Fuel Economy	燃油经济性
FP	Fuel Pump	燃油泵
FRP	Fuel Rail Pressure	燃油轨油压
FWD	Front Wheel Drive	前轮驱动
FWD/REV	Forward/Reverse	前进/后退
GEN	Generator	交流发电机
GND	Ground	搭铁,接地

HC	Hydrocarbon	碳氢化合物
HI	high	高的
HID	High Intensity Discharge	高压放电
IA	Intake Air	进气
IAB	Intake Air Bypass	进气歧管
IAC	Idle Air Control	怠速控制
IACV	Idle Air Control Valve	怠速空气控制阀
IAT	Intake Air Temperature	进气温度
IATS	Intake Air Temperature Sensor	进气温度传感器
ICE	Internal Combustion Engine	内燃机
ID(I.D.)	Inside Diameter	内径
IDI	Indirect Injection	(柴油机)间接喷射
IFI	Indirect Fuel Injection	(柴油机)间接喷射
IG(IGN)	Ignition	点火燃烧
IMA	Idle Mixture Adjustment	怠速混合气调整
IMPS	Intake Manifold Pressure Sensor	进气歧管压力传感器
I/P	Instrument Panel	仪表盘
ISC	Idle Speed Control	怠速控制
J/B	Junction Block	接线盒
KS	Knock Sensor	爆震传感器
LHD	Left Handle Drive	左侧驾驶
L	liter	升
L/C	Lock-up Clutch	锁死离合器
LCD	Liquid Crystal Display	液晶显示(器)
LED	Light Emitting Diode	发光二极管
LSD	Limited Slip Differential	防滑差速器
L-4	In-Line Four Cylinder	直列四缸
M/C	Mixture Control	混合气控制
MAF	Mass Air Flow	空气流量
MAF	Mass Air Flow Sensor	空气流量计
MAP	Manifold Absolute Pressure	歧管绝对压力
MAPS	Manifold Absolute Pressure Sensor	歧管绝对压力传感器
MAT	Manifold Air Temperature	歧管空气温度
MAX	maximum	最大
MCK	Motor Check	马达检测
MCS	Mixture Control Solenoid	混合气控制电磁线圈
MCU	Microprocessor Control Unit	微处理器控制单元
MF	Maintenance Free	免维护

MIL	Malfunction Indicator Lamp	故障指示灯
MIN	minimum	最小
M/S	Manual Steering	手动(机械)转向
M/T	Manual Transmission	手动变速器
NPS	Neutral Position Switch	空挡开关
OD	Outside Diameter	外径
OE	Opposed-cylinder Engine	对置发动机
O/D	Overdrive	超速挡
O/D off	Overdrive off	脱开超速挡
OH	Overhaul	大修
O-H	Off Highway	越野
OHC	Overhead Camshaft	顶置凸轮轴
OHV	Overhead Valve	顶置气门
OL, OP	Open Loop	开环
OS	Oxygen Sensor	氧传感器
P	Park	停车
P/N(PNP)	Park/Neutral Position	停车/空挡位置
P/S	Power Steering Pressure Switch	动力转向压力开关
PCM	Powertrain Control Module	动力控制模块
PCV	Positive Crankcase Ventilation	曲轴强制通风
PIP	Position Indicator Pulse	曲轴位置传感器
PKB	Parking Brake	驻车制动器
PMR	Pump Motor Relay	油泵马达继电器
PRV	Pressure Relief Valve	卸压阀
PSF	Power Steering Fluid	动力转向油
PSP	Power Steering Pressure	动力转向压力
PSPS	Power Steering Pressure Switch	动力转向油压开关
PSW	Pressure Switch	压力开关
PWR	Power	电源
REF	Reference	参照,基准
REV	Reverse	倒挡
RHD	Right Handle Drive	右侧驾驶
ROM	Read only Memory	只读存储器
RPM	Revolution Per Minute	转数每分钟
RR	Rear Engine Reardrive	后置发动机后轮驱动
RTS	Rapid Transit System	快速运输系统
RWB	Rear Wheel Brake	后轮制动
RWD	Rear Wheel Drive	后轮驱动

RWS	Rear Wheel Steering	后轮转向
SAE	Society of Automotive Engineers	美国汽车工程师学会
SBEC	Single Board Engine Control	单板发动机控制
SC	Supercharge	增压器
SCV	Swirl Control Valve	湍（涡）流控制阀
SMEC	Single Module Engine Control	单片发动机控制
SOHC	Single Overhead Camshaft	顶置单凸轮轴
SOL	Solenoid	线圈
SPD	Speed	车速,转速
SRS	Supplemental Restraint System	安全气囊
ST	start	起动
T/A	Transaxle	变速驱动桥
TAC	Tachometer	转速表
TC	Turbocharger	涡轮增压器
TCC	Torque Converter Clutch	变矩器离合器
TCCP	Torque Converter Clutch Pressure	变矩器离合器液压
TCM	Transmission Control Module	变速器控制模块
TCS	Traction Control System	牵引力控制系统
TCV	Timing Control Valve	正时控制阀
TD	Turbo Diesel	涡轮增压柴油机
TDC	Top Dead Center	上止点
TDCL	Test Diagnostic Communication Link	自诊接头
TDI	Turbo Direct- Injection	涡轮增压直接喷射
TEMP	Temperature	温度
T/F	Transfer Box	变速器
TME	Transverse Mid-engine	横向中置发动机
T/M	Transmission	变速器
TPC	Tire Performance Criteria	轮胎性能指标
TPNP	Transmission Park Neutral Position	变速器驻车空挡位置
TRC	Traction Control	牵引控制
TRK	Truck	载货汽车
TRS	Transmission Range Selection	变速器挡位选择
TS	Transmission System	传动系统
TURBO	Turbocharge	涡轮增压
TVBV	Turbocharger Vacuum Bleed Valve	涡轮增压器真空放气阀
TWC	Three Way Catalytic Converter	三效催化转化器
TWD	Two Wheel Drive	两轮驱动
TWILT	Twilight	双灯

UB	Under Body	车身底部
UDC	Upper Dead Center	上止点
UI	Unit Injector	泵—喷油器
UJ	Universal Joint	万向节
ULE	Ultra -low Emission	超低排放
V6		V 形 6 缸发动机
VAC	Vacuum	真空
VAF	Volume Air Flow	空气流量
VAT	Vane Air Temperature	进气温度
VC	Viscous Clutch	液力离合器
VCC	Viscous Converter Clutch	变矩离合器
VCS	Vacuum Control Solenoid	真空控制阀
VCV	Vacuum Control Valve	真空控制阀
	Vacuum Check Valve	真空止回阀
VR	Voltage Regulator	调压器
VSS	Vehicle Speed Sensor	车速传感器
VSV	Vacuum Solenoid Valve	真空电磁阀
VC	Viscous Coupling	液压耦合
WS	Wheel Steering	车轮转向
WSS	Wheel Speed Sensor	车速传感器
ZEV	Zero Emission Vehicle	零污染排放车辆

参 考 文 献

[1] 李万莉.工程机械专业英语[M].北京:人民交通出版社,1998.
[2] 宋永刚.工程机械专业英语[M].北京:人民交通出版社,2006.
[3] 王静文.汽车工程专业英语[M].北京:人民交通出版社,1999.
[4] 王怡民.汽车专业英语[M].北京:人民交通出版社,2003.
[5] 黄韶炯.汽车专业英语[M].北京:人民交通出版社,2005.
[6] 李自光,展朝勇.公路施工机械[M].北京:人民交通出版社,2005.
[7] 贾长海,展朝勇,郑忠敏.公路养护机械与养护机械化[M].北京:人民交通出版社,2004.